T0131046

RAND's Scalable Warning and Resilience Model (SWARM)

Enhancing Defenders' Predictive Power
in Cyberspace

BILYANA LILLY, ADAM S. MOORE, QUENTIN E. HODGSON,
DANIEL WEISHOFF

Prepared for the Office of the Secretary of Defense
Approved for public release; distribution unlimited

RAND NATIONAL DEFENSE RESEARCH INSTITUTE

For more information on this publication, visit www.rand.org/t/RRA382-1

Library of Congress Cataloging-in-Publication Data is available for this publication.
ISBN: 978-1-9774-0677-4

Published by the RAND Corporation, Santa Monica, Calif.
© 2021 RAND Corporation
RAND® is a registered trademark.

Cover image: Liuzishan/Getty Images

Support RAND
Make a tax-deductible charitable contribution at
www.rand.org/giving/contribute

www.rand.org

Preface

In a networked information space characterized by a constantly evolving cyber threat environment, technology landscape, and attack surface, network defenders face the perpetual challenge of effectively preventing, detecting, and responding to cyber incidents. Although most organizations continuously strive for effective defense architectures, prediction of cyber incidents has remained elusive, as has the creation of an interrelated strategy for resilience prioritized by specific threats targeting a given organization. To help enable defenders to proactively protect their systems through early warning of cyber incidents before they occur, in this report we propose a cyber indications and warning–based model: the RAND Corporation's Scalable Warning and Resilience Model (SWARM). This model focuses on proactively addressing the nation-state–sponsored threat class by providing a practical method for defenders to identify threats to themselves, make a probabilistic prediction of when an attack is likely to occur, and improve resilience. We demonstrate the predictive power of this model in a case study.

This report should be of interest to chief information security officers and network defenders who are actively involved in defending their unclassified networks against the cyber espionage subclass of the state-sponsored activity threat class. Cyber-security researchers and cyber threat intelligence analysts will also benefit from the findings of this study.

The research reported here was completed in August 2020 and underwent security review with the sponsor and the Defense Office of Prepublication and Security Review before public release. This research was sponsored by the Office of the Secretary of Defense and conducted within the Cyber and Intelligence Policy Center of the RAND National Security Research Division (NSRD), which operates the National Defense Research Institute (NDRI), a federally funded research and development center (FFRDC) sponsored by the Office of the Secretary of Defense, the Joint Staff, the Unified Combatant Commands, the Navy, the Marine Corps, the defense agencies, and the defense intelligence enterprise. For more information on the Cyber and Intelligence Policy Center, see www.rand.org/nsrd/intel or contact the director (contact information is provided on the webpage).

Contents

Figures and Tables

Figures

Tables

Summary

Despite increased focus on developing more-sophisticated cybersecurity tools and techniques for defending organizations against cyber threats—including using cyber threat modeling, information sharing, and threat-hunting—the cyber defense community is still largely reacting to, rather than predicting or anticipating, these threats. The majority of cyber incidents are detected and addressed at the point of breach or after the cyber adversary has already infiltrated the information environment of an organization. This forces defenders to emphasize post-breach detection rather than prediction. In addition, the high volume of malicious activity alerts, threat data feeds, and indicator lists that defenders have to process as a part of their daily routine leads to work overload, which, in turn, causes exhaustion and a reduced attention to detail, decreasing the probability of effectively mitigating cyber incidents when they occur.

In this report, we endeavor to improve the defense strategy of cybersecurity teams by developing a practical and adaptable process-based approach, called the Scalable Warning and Resilience Model (SWARM). SWARM, described in Figure S.1, consists of four steps. It prioritizes threat detection for cyber intrusion sets most likely to target an organization's network, facilitates better prediction of cyber incidents, and enhances network resilience, without assuming that an organization has access to classified information or assets.

SWARM adapts the concept of resilience and indications and warning frameworks, developed by the U.S. intelligence community, to information environments while also incorporating a combination of tailored threat modeling and emulation. The model allows defenders to prioritize their resources and focus on protecting their networks against the threat classes and adversaries that are most likely to target them based on their organization type. The model proposes an all-source intelligence collection process that includes both cyber threat intelligence (CTI) and strategic nontechnical data. Combined with the use of threat modeling frameworks, specifically PRE-ATT&CK and ATT&CK (adversarial tactics, techniques, and common knowledge),[1] the model recommends building a comprehensive profile of the cyber adversaries that

---

[1] As of October 2020, MITRE has removed the PRE-ATT&CK domain and integrated two new techniques into the enterprise ATT&CK framework. See MITRE | ATT&CK, "Updates," webpage, last modified April 29, 2021.

Figure S.1
RAND's Four-Step Scalable Warning and Resilience Model

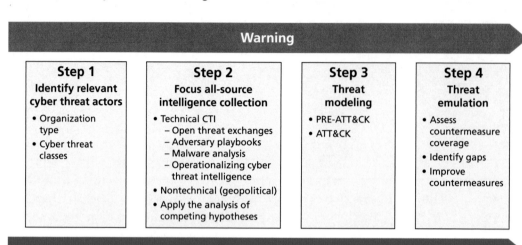

an organization is likely to face. In the final step, the model advocates periodic red and purple team exercises during which defenders test their systems, assess countermeasure coverage, identify gaps, and update their defenses based on the results of the performed exercises.

SWARM is meant to achieve two major objectives. First, it is designed to increase predictive power by providing advance warning for cyber incidents through early and more-comprehensive indicators, both technical and nontechnical. Second, the framework intends to enhance network resilience against targeted cyber incidents. This model should not be an organization's sole defense mechanism but should be one component of an effective layered defensive cyber operations approach.

This report is a result of the combination of several methodological techniques. The literature consulted in this analysis was compiled through a systematic literature review covering academic literature, reports, and statements from cybersecurity practitioners in the military and civil agencies of the United States, other governments, and international organizations.

To illustrate how to apply SWARM, we present a case study based on a series of unsuccessful North Korean cyber intrusions that targeted the RAND Corporation's information environment in 2018 and 2019. This case study uses data provided by the RAND Cyber Defense Center and applies SWARM to explore whether there are meaningful correlations between certain geopolitical events and subsequent phishing campaigns targeting RAND that are attributed to the North Korean intrusion set, known alternatively as Kimsuky, Thallium, or Velvet Chollima by cybersecurity companies Kaspersky, Microsoft, and CrowdStrike, respectively.

Considering the types of intrusion sets that target RAND and other targets, cyber intrusion attempts against RAND seem to be initiated primarily because of the perception of RAND as a think tank. State-sponsored espionage and cybercrime threat classes both target RAND. The identification of threat actors in Step 1 answers the question, Who is targeting or most likely to target one's information environment? This information can assist with the prioritization of technical CTI collection.

Step 2 provides an overview of the basic technical CTI collection and operationalization process used to enhance resilience, as well as strategic-level threat intelligence in the form of nontechnical open-source intelligence analysis to answer the questions: Why are attackers targeting you? What are their goals? What are their objectives and what environmental triggers may exist to indicate an imminent attack or campaign? More than likely, reviewing open-source intelligence and closed-source information, including one's own incident data and additional analysis performed while operationalizing tactical indicators of compromise, will provide some answers to these questions to proceed with predictive analysis.

Reports reviewed also yielded techniques, tactics, and procedures useful for modeling the threat in Step 3 beyond what had been observed in previous campaigns against RAND. This provided an expanded understanding of the adversary's behavioral playbook, which can then be used to emulate Kimsuky's techniques against one's own organization in Step 4 to test and improve prevention, detection, and countermeasures; discover visibility gaps; and continue to further enhance resilience against future attacks. Our analysis concluded that there seems to be a correlation and possible causation of Kimsuky phishing campaigns that stems from sanctions-related geopolitical events. More data are desired, especially related to phishing campaigns; however, we must wait for more phishing campaigns to introduce more data points.

SWARM presents a methodological process-based approach for a robust defense architecture that will facilitate prediction and early warning of cyber incidents relative to one's information environment while enhancing resilience, which is especially critical for organizations in the rapidly evolving cyber threat landscape. As cyber defense teams implement and adapt SWARM for their organizations, we believe improvements and refinement will result from this operational experience.

Acknowledgments

We would like to acknowledge the invaluable feedback on components of this work provided by Caolionn O'Connell, Osonde Osoba, Lillian Ablon, Jair Aguirre, Jonathan Fujiwara, and Andrew Lohn. We thank Eric Kao for some of the threat-modeling work. Numerous representatives of the private sector, the U.S. government, and international organizations also shared unique insights that inspired and shaped different elements of the arguments outlined in this report. We are immensely grateful for their patience and willingness to discuss this sensitive issue. We also thank Neil Jenkins and Sasha Romanosky, who provided valuable feedback on the draft report. We thank Rich Girven and Sina Beaghley for their guidance and support for this project. We are grateful to the organizers of the North Atlantic Treaty Organization International Conference on Cyber Conflict in May 2019, where our lead author presented precursor research that informed this report, and to the William and Flora Hewlett Foundation, for hosting a policy salon discussion at the hacker conference DEFCON in August 2019, where we engaged in discussions and received feedback on our initial thinking regarding a warning and resilience model, which informed the final model proposed in this research.

Abbreviations

ACH	analysis of competing hypotheses
APT	advanced persistent threat
ATT&CK	adversary tactics, techniques, and common knowledge
C2	command and control
CERT	computer emergency response team
CISO	chief information security officer
CLI	command line interface
CTI	cyber threat intelligence
CVE	common vulnerabilities and exposures
DDoS	distributed denial of service
DHS	U.S. Department of Homeland Security
DIA	Defense Intelligence Agency
DoD	U.S. Department of Defense
DNI	Director of National Intelligence
DNS	Domain Name System
EDR	endpoint detection and response
I&W	indications and warning
INSA	Intelligence and National Security Alliance
IOA	indicator of attack
IOC	indicator of compromise

IP	internet protocol
IR	incident response
LAMP	Lockwood Analytical Method for Prediction
MD5	Message Digest 5
MISP	Malware Information Sharing Platform
NATO	North Atlantic Treaty Organization
NGO	nongovernmental organization
NIST	National Institute of Standards and Technology
OSINT	open-source intelligence
OTX	Open Threat Exchange
pDNS	Passive Domain Name System
PIRs	priority intelligence requirements
PMESII	political, military, economic, social, information, and infrastructure
RGB	Reconnaissance General Bureau (North Korea)
SIEM	security information and event manager
SSL	Secure Sockets Layer
SWARM	Scalable Warning and Resilience Model
TTPs	tactics, techniques, and procedures
UN	United Nations
URL	uniform resource locator
USCERT	U.S. Computer Emergency Response Team
USCYBERCOM	U.S. Cyber Command
USIC	U.S. Intelligence Community
VBA	Visual Basic for Applications
VT	VirusTotal

CHAPTER ONE

Introduction, Research Methodology, and Historical Evolution of Concepts

The evolving cyber threat landscape requires rapidly and constantly adapting cyber defense solutions. The current strategies that defenders employ are based predominantly on detecting cyber incidents at the early or later stages of a cyberattack cycle but seldom prior to the delivery of a weaponized payload to the defenders' networks. Other initial stages of a cyberattack cycle—the adversary conducting reconnaissance on the target, testing capabilities, establishing and maintaining infrastructure, or potential geopolitical trigger events occurring—are not typically factored into defenders' calculations when predicting or preventing cyber incidents. As a result, the model introduced in this report intends to enhance the predictive and anticipatory capabilities available to cyber defenders while also augmenting resilience by improving preventions and detections as early in Lockheed Martin's Cyber Kill Chain framework as possible.[1]

Aware of these limitations, government and private-sector cybersecurity experts have expressed discontent with the current nature of cyber defenses, suggesting their willingness to transform their cybersecurity strategies from reactive to proactive. Such a transformation should include a strategy to predict and tailor preparation against cyber threats before such threats establish contact with defenders' networks and would allow for the implementation of faster and more-efficient preventive and mitigating policies.

An emerging body of literature on data collection and predictive analytics can facilitate the incorporation of information from these early stages of cyberattack cycles to help defenders anticipate and detect attacks with a certain level of probability before cyber threats establish contact with targeted networks. Yet, to date, there is insufficient analysis of how these methods can be applied systematically to an information environment to facilitate the detection of cyber activities in the early stages of cyber incidents.

---

[1] The Lockheed Martin Cyber Kill Chain is a framework for mapping the seven common stages of cyber incidents in support of intelligence-driven defense. For more information, see Michael Muckin and Scott C. Fitch, *A Threat-Driven Approach to Cyber Security Methodologies: Practices and Tools to Enable a Functionally Integrated Cyber Security Organization*, Bethesda, Md.: Lockheed Martin Corporation, 2019.

Research Methodology and Report Organization

This report proposes addressing this issue and facilitating the formulation of a proactive defense strategy against cyber incidents based on the application of the concepts of both warning *and* resilience with the implementation of a Scalable Warning and Resilience Model (SWARM) for cyberspace. SWARM demonstrates the methodical integration of recently developed analytic techniques and tools for data collection and analysis that can assist defenders in identifying threats to their information environment in advance of an incident. The proposed high-level framework is characterized as scalable because it is adaptable and can be applied and integrated within the existing capabilities of any organization across governments, the private sector, and international organizations.

The development of SWARM is based first on an analysis of publicly available mature indications and warning (I&W) frameworks used by the U.S. intelligence community (USIC). These particular USIC frameworks provide a rigorous methodological foundation for a SWARM because of their extensive and repeatedly tested application. Furthermore, there is sufficient publicly available literature on these frameworks' components that enables the formulation of a comprehensive understanding of the frameworks, their advantages, and their pitfalls.

The proposed SWARM framework outlined in this report is based solely on publicly available open-source materials. Because of the highly sensitive nature of this topic, which results in a limited capacity to examine all aspects of this issue in depth, the arguments put forth in this analysis are suggestive, not definitive. They are intended to serve as a foundation and starting point, presenting a list of initial options that defenders can adopt and adapt to their cyber defense programs.

This report is a result of the combination of several methodological techniques. The literature consulted in this analysis was compiled through a systematic literature review of relevant databases and search engines, including JSTOR, EBSCO, Applied Science and Technology Full Text, Conference Proceedings Citation Index—Science, Conference Proceedings Citation Index—Social Sciences and Humanities, Google, Google Scholar, ACM Digital Library, IEEE Xplore, and Web of Science. The literature review also encompassed official documents and statements from cybersecurity practitioners in the military and civil agencies of the United States, other governments, and international organizations, such as the North Atlantic Treaty Organization (NATO). We reviewed national cyber and military doctrines and media articles. The findings of this report were further informed and shaped by a series of interviews that the research team conducted with current and former U.S. government representatives, cyber experts working within NATO structures, members of the cybersecurity private sector, and cybersecurity researchers. Through the use of a *snowballing convenience sampling*, where researchers identify interviewees who were recommended by previously interviewed experts, the researchers identified additional individuals and organizations to consult and further enrich and refine the arguments proposed in this study.

Because the focus of the proposed model is on both warning *and* resilience, we first describe the unclassified history and evolution of the I&W concept in the USIC. We then examine the nature and history of the concept of resilience and its application to cyberspace. We proceed by briefly reviewing the main terminology related to the concepts discussed in this report.

After outlining the major concepts used in this report and defining warning, resilience, and other key terms, we examine several classic I&W frameworks applied by the USIC. We highlight these frameworks' common stages and the different emphasis in each framework to create a list of priority elements used to inform a practical warning and resilience model. The analysis then reviews how resilience has been applied to the cyber domain. On the basis of the analysis of these traditional I&W frameworks and the concept of resilience, we outline a threat-based four-step model that can be applied to a defenders' information environment to facilitate the anticipation and early detection of cyber incidents and to enhance resilience.[2] We proceed by examining, in detail, each stage of the model and identifying recent data collection and analysis techniques and tools that can be integrated in the various stages of the framework. After outlining the different steps of the model, we conclude by demonstrating the application and potential effectiveness of the SWARM framework based on a case study of a cyber incident that occurred at the RAND Corporation.

This model would ideally be implemented as a part of a broader cyber defense strategy that an organization applies to its information environment. Such strategy should include identification and systematic monitoring of the organization's network and key assets prior to the implementation of this model. The implementation of functional cyber hygiene activities, such as vulnerability management, training, and controlling access, can further enhance organizational resilience.[3]

History and Evolution of Indications and Warning Frameworks in the USIC

Warning intelligence, also referred to as indications intelligence or I&W, is primarily a product of the post–World War II period. Concerned that potential adversaries would undertake an attack similar to the Japanese attack on Pearl Harbor in 1941 without a

---

[2] *Cyber incidents* are "actions taken through the use of computer networks that result in a compromise or an actual or potentially adverse effect on an information system and/or the information residing therein." See U.S. Code of Federal Regulations, Title 48, Chapter 2, Subchapter H, Part 252, Subpart 252.2, Section 252.204-7012, Safeguarding Covered Defense Information and Cyber Incident Reporting.

[3] Matthew Trevors, "Mapping Cyber Hygiene to the NIST Cybersecurity Framework," webpage, Software Engineering Institute, Carnegie Mellon University, October 30, 2019.

formal declaration of war or other warning, the USIC began developing robust methodologies for evaluating and continuously monitoring potential adversaries' activities. These methodologies were attempts to anticipate similar threats with sufficient time to avoid surprise attacks and implement countermeasures.[4]

There are several discernible periods in the evolution of the warning intelligence analytic discipline since its inception, during which the major threats faced by the United States and its intelligence community redefined the nature of the warning problem. The first is the Cold War period, when warning intelligence was focused on existential threats to the United States posed by the Soviet Union. This was followed by the period of the 1990s to 2000s, when the focus on I&W expanded to include non-state actors, and then, finally, by the period from 2007 onward, when cyber capabilities and cyber operations became tools of statecraft and necessitated an adaptation of I&W frameworks to cyberspace.

Since around 1948, warning methodologies were mainly focused to provide timely warning of "the threat of possible military action by a foreign power."[5] The primary focus was, naturally, on the capabilities and intentions of the Soviet Union and the other members of the Sino-Soviet bloc.[6] In the 1940s, the threat from the Soviet Union and the countries it controlled in Eastern Europe—and the threat of communist China starting in 1949—was treated as monolithic. Therefore, the indicator lists used to monitor potential threats from these areas were generally applied to all communist states without being refined to address specific developments and unique characteristics of each potential adversary. In the 1950s, the intelligence community began to design and adopt more country- and region-specific lists that more accurately captured developments in the Taiwan Strait, Southeast Asia, and other locations.[7]

After the collapse of the Soviet Union, and especially after the attacks of September 11, 2001, terrorism emerged as a major threat to the security of the United States and its allies. This development shifted the focus on I&W and expanded the targets of data collection and analysis to incorporate the analysis of capabilities and intentions of nonstate actors (in addition to states) that can pose a threat to U.S. security interests and the interests of U.S. partners.[8] The 2007 campaign of distributed denial of service (DDoS) attacks against Estonia's government and critical infrastructure added cyber capabilities and operations to the tools of statecraft and marked another milestone in the history of I&W frameworks. Since the 2007 attacks, detected activities and threats

---

[4] Cynthia M. Grabo, *Anticipating Surprise: Analysis for Strategic Warning*, Jan Goldman, ed., Center for Strategic Intelligence Research, Joint Military Intelligence College, 2002, p. 1.

[5] Cynthia M. Grabo, *Warning Intelligence*, McLean, Va.: Association of Former Intelligence Officers, 1987, p. 3; Grabo, 2002, pp. ix, 25.

[6] Grabo, 2002, pp. v, 25.

[7] Grabo, 2002, p. 25.

[8] Grabo, 2002, p. ix.

in cyberspace have considerably increased in magnitude and frequency, ranging from international ransomware campaigns to attacks on electrical grid infrastructure that can cause physical damage, such as the 2015 attack on a Ukrainian electricity distribution company.[9] The challenges in cyberspace illustrate the need to apply I&W frameworks to network security.

The current wide spectrum of actors, methods, and scenarios that can pose a risk to U.S. and allied interests is reflected in a broader definition of threats, including any "discernible danger" that can inflict potential damage "to U.S. or allied persons, property or interests that occurs in a definable time in the future."[10] Such threats include hostile actions by states, conflicts affecting U.S. security interests, and terrorist activity.[11] This broad definition allows for monitoring threats in the traditional domains of land, sea, air, and space, as well as in cyberspace.

Main Definitions of I&W Frameworks

An examination of national and international strategies suggests that, despite the importance of the I&W concept, there is no formally accepted definition of I&W among the USIC, the cyber community, and U.S. partners and allies. Various governments and even departments within governments have adopted slightly different interpretations of I&W frameworks. Similarly, related terminology, such as *indicator* and *indication*, also lack common definitions and have competing interpretations in different communities. This conceptual ambiguity prevents integrating and applying common I&W frameworks across government departments and domains. To avoid confusion, bridge the terminological divide, and establish a common framework of understanding, this section defines some of the conceptual variations of the key terms related to I&W and resilience, offers a definition of cyber I&W, and proposes an argument for the appropriateness of integrating I&W and resilience in cyberspace.

I&W is a well-established concept and theory of practice in the USIC and U.S. military. I&W is used to notify decisionmakers of adversarial activities that could have a negative impact on security interests or military forces.[12] In the U.S. Department of Defense (DoD) lexicon, I&W is an intelligence activity "intended to detect and report time-sensitive intelligence information on foreign developments that forewarn

---

[9] Robert M. Lee, Michael J. Assante, and Tim Conway, *Analysis of the Cyber Attack on the Ukrainian Power Grid*, Washington, D.C.: SANS Industrial Control Systems, Electricity Information Sharing and Analysis Center, March 18, 2016.

[10] Defense Intelligence Agency, "Warning Fundamentals," unclassified briefing, May 2014, slide 4.

[11] Grabo, 2002, p. 2; Intelligence and National Security Alliance, *A Framework for Cyber Indications and Warning*, Arlington, Va., October 2018, p. 3.

[12] Bruce W. Watson, Susan M. Watson, and Gerald W. Hopple, eds., *United States Intelligence: An Encyclopedia*, New York: Garland Publishing, 1990, p. 594; Grabo, 1987, p. 5.

of hostile actions or intention against United States entities, partners, or interests."[13] It is an analytical process providing a structured way of continuously monitoring, reporting on, and detecting developments that would suggest the occurrence of a potential threat. The I&W process uses political, military, and economic events—as well as associated developments and actions that could shed light on potential preparations for hostile activities—to build a probabilistic assessment. The process is rooted in the art of probabilities providing a qualified (high, medium, or low) or definitive (positive or negative) evaluation of the likelihood of a particular threat. The produced intelligence is actionable, which signifies that it can be brought to the attention of decisionmakers and used to inform timely policy.[14]

DIA defines *I&W* as "[a] distinct communication to a decision maker about threats against U.S. and allied security, military, political, information, or economic interests. The message should be given in sufficient time to provide the decision maker opportunities to avoid or mitigate the impact of the threat."[15]

The contours of a *cyber* I&W concept are not well established in the cybersecurity community. The cyber community accepts that I&W is a structured process, but cyber schools of thought diverge on the type of collected information and the temporal scope of the information that should be included in a cyber I&W framework. Some experts and practitioners think cyber I&W should include both technical and geopolitical data, while others contend that I&W processes should analyze only technical data. Most of our current defenses are built primarily using technical data. Furthermore, some experts consider cyber I&W to include monitoring adversary's activities only until the delivery stage of a weaponized payload, which is the third stage of the Cyber Kill Chain. Still others think adversarial actions past the delivery stage should also be included.[16]

To standardize our understanding, for the purposes of this analysis, we created our own definition of *cyber I&W* or *cyber warning intelligence*, which is

> an analytical process focused on collecting and analyzing information from a broad array of sources to develop indicators which can facilitate the prediction, early detection, and warning of cyber incidents relative to one's information environment.[17]

---

[13] Joint Publication 2-0, *Joint Intelligence*, Washington, D.C.: Office of the Joint Chiefs of Staff, October 22, 2013, p. GL-12.

[14] Grabo, 2002, pp. iii–3; Bilyana Lilly, Lillian Ablon, Quentin Hodgson, and Adam Moore, "Applying Indications and Warning Frameworks to Cyber Incidents," in Tomáš Minárik, Siim Alatalu, Stefano Biondi, Massimiliano Signoretti, Ihsan Tolga, and Gábor Visky, eds., *11th International Conference on Cyber Conflict: Silent Battle, Proceedings 2019*, Tallinn: NATO Cooperative Cyber Defence Centre of Excellence, 2019, p. 84.

[15] Quoted in DIA, undated.

[16] Lilly et al., 2019, pp. 85–86.

[17] Lilly et al., 2019, p. 85.

The definition does not specify what type of data should be collected in this analysis, allowing for the inclusion of both technical and geopolitical data. The inclusion of geopolitical indicators can provide valuable information on environmental triggers that can facilitate the anticipation of an incident. The USIC has learned the advantages of such comprehensive approaches to data collection and analysis. Cynthia Grabo, an eminent threat intelligence analyst, notes that a warning methodology should include both political and military indicators.[18] Knowledge of precedent, history, and doctrine is relevant and necessary for the creation of attack scenarios and for the design of indicators, which typically originate from three sources: historical precedent, adversary behavior during recent crisis, and knowledge of adversarial playbooks and doctrine.[19] These data collection domains and the respective methods applied to analyzing the gathered information are applicable to the identification and profiling of cyber threats, which is expounded further in the subsequent sections of this report. Furthermore, this definition does not specify which stages of a cyberattack cycle should be included in an I&W framework. Cyber I&W frameworks should cover both pre- and postdelivery stages of cyber incidents, with a focus on the predelivery stages, especially reconnaissance, to allow for the development of proactive defense that harnesses the power of predictive analytics.

In addition to variance in definitions of I&W, the intelligence and the cyber communities have different definitions of the terms *indicator* and *indication*. The difference between these two concepts in the USIC is one between theory and practice, or between expectation and an actual occurrence. An *indicator* refers to an evidence-based analytical judgment used to identify "theoretical or known development or an action which the adversary may undertake in preparation for a threatening act such as a deployment of forces, a military alert, a call-up of reservists, or the dispatch of a diplomatic communique."[20] The intelligence community adds such indicators to a list that it continuously monitors, known as an *indicator list*. A signal that a monitored *indicator* is taking place in reality is referred to as an *indication*. Such *indication* is used as a part of an assessment on potential adversarial actions.[21]

In contrast to the USIC, the cybersecurity community uses the term *indicator of compromise* (IOC) to refer to evidence on a computer or in transit over a network that signifies a breach in the security of a network. Such evidence could include a command and control (C2) domain address, a Message Digest 5 (MD5) hash, or a file name. This evidence is typically gathered after a compromise to a system has already occurred.[22]

---

[18] Grabo, 1987.

[19] Grabo, 2002, pp. 13, 26.

[20] Grabo, 2002, p. 3; Lilly et al., 2019, p. 84.

[21] Grabo, 2002, p. 3; Lilly et al., 2019, p. 84.

[22] Jessica DeCianno, "IOC Security: Indicators of Attack vs. Indicators of Compromise," blog post, *Crowd-Strike*, December 9, 2014.

Although IOCs are then integrated into defense systems and are added to deny lists and intrusion detection systems to facilitate prevention and detection, they still primarily detect offensive cyber operations during the delivery stage or later in Lockheed Martin's Cyber Kill Chain. In the cybersecurity context, the use of IOC resembles the use of the term *indication*, not *indicator*, in the intelligence community. In this report, we use the intelligence community definitions of the terms *indicator* and *indication*, while we refer to IOC using the definition adopted in the cybersecurity community.

A related term in the cybersecurity community is *indicators of attack* (IOAs). It focuses on detecting the intention of a potential cyber adversary and the outcome that the adversary is trying to accomplish. IOAs can be incorporated into the design of more-proactive anticipatory defenses.[23]

History and Evolution of the Concept of Resilience

Resilience is a term used across many disciplines, often with variations in its meaning. It appears in literature related to ecology, sociology, and engineering, and has gained particular prominence in recent years in the context of critical infrastructure protection.[24] In ecology, *resiliency* is related to notions of adaptability to changing environments and the emergence of new threats, while political science literature speaks more to the maintenance of mission effectiveness in the face of threats.[25] The Electric Power Research Institute provides three elements in its definition of *grid resiliency*: prevention, recovery, and survivability.[26] This is similar to the approach that the U.S. Department of Homeland Security's (DHS's) National Infrastructure Advisory Council took in its 2009 report, *Critical Infrastructure Resilience*, which defined *infrastructure resilience* as "the ability to reduce the magnitude and/or duration of disruptive events. The effectiveness of a resilient infrastructure or enterprise depends upon its ability to anticipate, absorb, adapt to, and/or rapidly recover from a potentially disruptive event."[27] These reports and studies all make an attempt to define resilience, while many others use the term without defining it, ignoring its multiple extant definitions and uses. Others take the standard definition in dictionaries, such as *Merriam-Webster's Unabridged Diction-*

---

[23] DeCianno, 2014.

[24] Barack H. Obama, Executive Order 13636, *Improving Critical Infrastructure Cybersecurity*, Washington, D.C.: White House, February 12, 2013.

[25] Krista S. Langeland, David Manheim, Gary McLeod, and George Nacouzi, *How Civil Institutions Build Resilience: Organizational Practices Derived from Academic Literature and Case Studies*, Santa Monica, Calif.: RAND Corporation, RR-1246-AF, 2016.

[26] Electric Power Research Institute, "Grid Resiliency," webpage, undated.

[27] National Infrastructure Advisory Council, *Critical Infrastructure Resilience: Final Report and Recommendations*, Washington, D.C.: U.S. Department of Homeland Security, September 8, 2009, p. 8.

ary, which states that *resilience* is "the ability to bounce or spring back into shape, position, etc., after being pressed or stretched."[28] This usage clearly refers to the resilience of a physical material, rather than complex systems, because a network is not usually subject to literal pressing or stretching. T. D. O'Rourke, an engineering professor, uses engineering concepts developed at the Multidisciplinary Center for Earthquake Engineering Research to highlight the four qualities of infrastructure resilience: robustness, redundancy, resourcefulness, and rapidity (i.e., how quickly a system can recover). J. D. Taft argues that many of these definitions include notions of an ability to continue operations in the face of threats and the ability to recover quickly after system disruptions, and that these definitions are, in fact, conflating resilience and reliability. Taft was writing specifically in the context of electric grid resilience, but his argument could be extended to a general discussion of resilience in any system or network. He argues that resilience is "an intrinsic characteristic" of a system or network (more specifically a grid in his argument). He further argues that reliability is a measure of "behavior once resilience is broken."[29]

The concept of resilience has also been applied to the cyber domain for at least the past decade. Researchers at Carnegie Mellon tied operational resilience to the work of computer emergency response teams (CERTs) in 2010, highlighting that an organization's cybersecurity team's focus is tied to the organization's operations rather than to a general security function.[30] The MITRE Corporation's Deborah Bodeau and Richard Graubart developed a framework for cyber resiliency engineering, positing goals of (1) anticipating threats, (2) continuing essential functions in the face of even successful attacks, (3) restoring those functions, and (4) evolving the organization's business functions moving forward.[31] As with resilience in the other domains noted previously, this framework combines notions of continuing operations to the maximum extent possible and recovery. The National Institute of Standards and Technology (NIST) has adopted this approach as well, defining information system resilience as "the ability of an information system to continue to: (i) operate under adverse conditions or stress, even if in a degraded or debilitated state, while maintaining essential operational capabilities; and (ii) recover to an effective operational posture in a time frame consistent

---

[28] Quoted in T. D. O'Rourke, "Critical Infrastructure, Interdependencies, and Resilience," *The Bridge*, Vol. 37, No. 1, Spring 2007, p. 25.

[29] J. D. Taft, *Electric Grid Resilience and Reliability for Grid Architecture*, Richland, Wash.: Pacific Northwest National Laboratory, U.S. Department of Energy, PNNL-26623, November 2017, p. 3.

[30] Richard A. Caralli, Julia H. Allen, and David W. White, *CERT Resilience Management Model: A Maturity Model for Managing Operational Resilience*, Boston, Mass.: Addison-Wesley Professional, 2010.

[31] Deborah J. Bodeau and Richard Graubart, *Cyber Resiliency Engineering Framework*, Bedford, Mass.: MITRE Corporation, September 2011, p. iii.

with mission needs."[32] Building from the work prompted by then-President Barack Obama's Executive Order 13636, *Critical Infrastructure Cybersecurity*, a joint Director of National Intelligence (DNI)– and DHS–sponsored working group also defined *resilience* in cyberspace as "the ability to *adapt* to changing conditions and *prepare* for, *withstand*, and rapidly *recover* from disruption."[33] The working group stated that

> cyber resiliency is that attribute of a system that assures it continues to perform its mission-essential functions even during a cyber incident. For services that are mission-essential, or that require high or uninterrupted availability, cyber resiliency should be built into the design of systems that provide or support those services.[34]

Cyber resilience has gained currency internationally as well, not just in the United States. The World Economic Forum inaugurated a dialogue on cyber resilience in 2011, defining *cyber resilience* as "the ability of systems and organizations to withstand cyber events, measured by the combination of mean time to failure and mean time to recovery." Again, the focus is on continuing mission functions to the maximum extent possible, and quickly recovering to a more secure state.[35] Often, resilience is distinguished from cybersecurity by asserting that resilience is an anticipatory rather than a reactive approach. The World Economic Forum's notion is that "the overarching goal of a risk-based approach to cybersecurity is system resilience to survive and quickly recover from attacks and accidents" and "accept that failures will occur. . . . [T]he objective is to restore normal operations and ensure that assets and reputations are protected."[36] Daniel Dobrygowski, the head of corporate governance and trust at the World Economic Forum, adds an additional facet to resilience, asserting that it is a long-term effort, not a discrete set of actions. This also supports the notion of an anticipatory posture—not only with respect to near-term threats, but also to evaluate and

---

[32] NIST, *Security and Privacy Controls for Federal Information Systems and Organizations*, Gaithersburg, Md.: Joint Task Force Transformation Initiative, Special Publication 800-53, Revision 4, April 2013, p. B-11.

[33] DNI and DHS, *Cyber Resilience and Response: 2018 Public-Private Analytic Exchange Program*, Washington, D.C., undated, p. 5.

[34] DNI and DHS, undated, p. 6.

[35] World Economic Forum, *Partnering for Cyber Resilience: Risk and Responsibility in a Hyperconnected World–Principles and Guidelines*, Geneva, Switzerland, 2012, p. 14. Others also discuss this combination of objectives in resilience. See also Zachary A. Collier, Igor Linkov, Daniel DiMase, Steve Walters, Mark Tehranipoor, and James H. Lambert, "Cybersecurity Standards: Managing Risk and Creating Resilience," *Computer*, Vol. 47, No. 9, September 2014; and Darko Galinec and William Steingartner, "Combining Cybersecurity and Cyber Defense to Achieve Cyber Resilience," *IEEE 14th International Scientific Conference on Informatics*, Poprad, Slovakia, November 19, 2017.

[36] World Economic Forum, 2012, pp. 9, 4.

posture an organization for changes in technology and new threats in the future.[37] The focus on cyber resilience has naturally extended to the commercial sector as companies advertise cyber resilience as a service they can provide to other organizations, whether as a product or as a service.[38]

As we noted previously, some writers have disputed whether resilience should include both concepts: continuing operations while under attack and recovering quickly. However, most writers, scholars, and practitioners believe that both concepts are important for resilience.[39] For our purposes, we focus on how cyber I&W can posture an organization to anticipate cyber threats to maintain mission effectiveness as much as possible. This can occur either through focused hardening of networks given anticipated threats or implementing alternative concepts of operations or procedures, such as shifting resources to other, segmented parts of the network.[40] Recovery is clearly an important element of a comprehensive cybersecurity strategy and approach; an organization can potentially posture itself for quick recovery by using its cyber warning to limit the impact of a malicious actor's cyber activity. That said, our focus in this report is on anticipatory actions. Therefore, we use the following definition of *resilience*:

> The ability of an organization to maintain operational capability during and after an attack either by preventing the attack or by mitigating the effectiveness of the attack via rapid detection and response.

This definition is consistent with many others while allowing for the integration of the concept of resilience into a warning intelligence framework that focuses on the promotion of organizational capacity to predict, anticipate, and "fight through" cyber incidents.

---

[37] Daniel Dobrygowski, "Cyber Resilience: Everything You (Really) Need to Know," blog post, World Economic Forum, July 8, 2016.

[38] Abi Tyas Tunggal, "Cyber Resilience: What It Is and Why You Need It," blog post, UpGuard, updated May 18, 2020; and Accenture, *The Nature of Effective Defense: Shifting from Cybersecurity to Cyber Resilience*, Dublin, 2018.

[39] See also Robert K. Knake, "Building Resilience in the Fifth Domain," blog post, Council on Foreign Relations, July 16, 2019.

[40] *Cyber hardening* is a component of resilience and focuses on decreasing the attack surface of a system.

CHAPTER TWO

Indications and Warning Frameworks

There are several publicly available classic intelligence I&W frameworks that the USIC and military have developed and applied over the past few decades to monitor potentially threatening behavior of likely adversaries. These frameworks include the Lockwood Analytical Method for Prediction (LAMP), the Seven Phases of the Intelligence-Warning Process, and DoD's *Defense Warning Network Handbook*.[1] These tested frameworks provide a valuable initial road map for the design of a cyber I&W framework and have served as an inspiration for a number of recently published I&W frameworks adapted to cyberspace. Some of these modified frameworks include Robinson, Astrich, and Swanson's adaptation of Lockwood's LAMP method (2012) and the high-level framework published by the Intelligence and National Security Alliance (INSA), an intelligence and national security trade association.[2]

These frameworks consist of a series of steps that describe an analytical process of data collection, analysis, and reporting. They vary in the level of specificity and the number of steps they contain and in the emphasis on different stages of the process. Lockwood's method, for example, contains 12 steps and emphasizes the view of the future as a changing spectrum of probability-based scenarios. This method recommends identifying these scenarios, estimating and ranking their relative probabilities, and designing indicators that could signal that these scenarios are coming to fruition. In comparison, the DoD *Defense Warning Network Handbook* still acknowledges the significance of identifying future potential scenarios but places more emphasis on exploring options to mitigate the threat and communicate the warning.[3] The Seven

---

[1] Jai Singh, "The Lockwood Analytical Method for Prediction Within a Probabilistic Framework," *Journal of Strategic Security*, Vol. 6, No. 3, Fall 2013. Also see Lilly et al., 2019; and Sundri Khalsa, "The Intelligence Community Debate over Intuition Versus Structured Technique: Implications for Improving Intelligence Warning," *Journal of Conflict Studies*, Vol. 29, April 1, 2009.

[2] Michael Robinson, Craig Astrich, and Scott Swanson, "Cyber Threat Indications and Warning: Predict, Identify and Counter," *Small Wars Journal*, July 26, 2012; and INSA, 2018.

[3] Department of Defense Directive 3115.16, *The Defense Warning Network*, Washington, D.C.: U.S. Department of Defense, Incorporating Change 2, August 10, 2020. For more-detailed examination of each framework, see Lilly et al., 2019.

Phases of the Intelligence-Warning Process provides a more balanced step-by-step approach highlighting the main aspects outlined in the LAMP and DoD approaches, and elaborates further on the need to focus data collectors on existing intelligence gaps that should be used to refine the monitoring and conclusions of the intelligence-warning process.

Recently published I&W frameworks that have been adapted to the cyber domain contain structural elements similar to those of intelligence I&W frameworks and typically incorporate a cyber threat model used to standardize the data collection, reporting, and analysis of cyber incidents. The process proposed by Robinson and colleagues is based on the LAMP method and suggests the use of Lockheed Martin's Cyber Kill Chain as a tool through which to standardize and derive information about adversaries' courses of action. The INSA framework proposes a similar approach and further recommends the use of MITRE's adversary tactics, techniques, and common knowledge (ATT&CK) framework. ATT&CK is a threat-based taxonomy of adversarial playbooks that is developed to map real-world behavior and techniques through gathering detailed technical data on the specific tactics, techniques, and procedures (TTPs) of different adversaries. ATT&CK maps adversary post-exploitation techniques and is an expansion of Lockheed Martin's Cyber Kill Chain.[4] Each ATT&CK playbook is a post-exploitation threat model and is based on the assumption that an adversary may use the same playbook for each operation or change technique combinations over time. MITRE has also developed the PRE-ATT&CK model to identify common behavioral steps preceding the breach of a network.[5]

Another model that can be included in a stage of an I&W framework to derive specific technical data about adversarial behavior is the Diamond model, which was developed by the Center for Cyber Intelligence Analysis and Threat Research. The Diamond model is used to classify data on four main elements of groups of attacks, rather than individual attacks. These elements are *adversary, infrastructure, capability,* and *victim.*[6]

Despite the variation in emphasis on different stages and levels of specificity provided in the I&W process, the examined frameworks contain common stages that can be grouped into seven categories. These stages are (1) framing the question, (2) identifying threats to defend, (3) identifying assets to defend, (4) developing attack scenarios, (5) developing indicators, (6) monitoring the developed indicators, and (7) acting on them. Table 2.1. shows how the stages of each of the models fits within these steps.

---

[4]   Blake Strom, "ATT&CK 101," blog post, Medium, May 3, 2018; and MITRE Corporation, "Enterprise Matrix," webpage, last modified July 2, 2020g.

[5]   MITRE | ATT&CK, "PRE-ATT&CK Matrix," webpage, last modified November 4, 2019d.

[6]   Chris Pace, ed., *The Threat Intelligence Handbook: A Practical Guide for Security Teams to Unlocking the Power of Intelligence*, Annapolis, Md.: CyberEdge Group, 2018, pp. 69–70.

These models serve as a valuable methodological foundation for the design of a rigorous and practical cyber I&W framework. Despite their benefits, the publicly available literature about these frameworks does not provide sufficient emphasis on a number of critical components—especially regarding the level of specificity of each step, such as what type of data should be collected for the analysis, how the data should be analyzed, what methods should be used for analysis—and how the frameworks should be integrated into already-existing cyber defense structures. The frameworks also do not address a number of significant challenges of the cyber domain—for example, the difficulty of identifying critical assets to defend in a constantly evolving and diverse information environment. The next chapter integrates the main components of the frameworks outlined in this chapter and proposes a practical and flexible warning and resilience model that network defenders can adapt to their information environment to defend against current and emerging cyber threats.

Table 2.1
Comparison of Classic and Cyber Indications and Warning Frameworks

| General Actions | Classic I&W Frameworks | | Cyber I&W Frameworks | |
	Defense Warning Network Handbook	LAMP/Lockwood	The Seven Phases of the Intelligence-Warning Process	Robinson et al. (adopted from LAMP)	INSA
Framing questions		1. Define the intelligence question under consideration with sufficient specificity and narrowness of enquiry		1. Problem identification: determine the issue	1. Develop a refined understanding of the most likely threats
Identify threats		2. Specify the actors involved in the program		2. Identify potential actors	2. Identify and prioritize assets to be protected
Identify assets to defend					3. Using structured analytic techniques, forecast likely attack scenarios (include Cyber Kill Chain and ATT&CK)

Table 2.1—Continued

General Actions	Classic I&W Frameworks			Cyber I&W Frameworks	
	Defense Warning Network Handbook	LAMP/Lockwood	The Seven Phases of the Intelligence-Warning Process	Robinson et al. (adopted from LAMP)	INSA
Develop scenarios	1. Identify anomalies/imagine alternatives	3. Study each actor's intentions and perceptions of the problem		3. Actor courses of action: Viability and probability (include the Cyber Kill Chain)	4. Decompose scenarios into indicators of likely adversary actions
	2. Produce scenarios	4. Specify all possible courses of action for each actor		4. Determine scenario enablement	
		5. Determine the major scenarios		5. Manifested scenario focal events	
		6. Calculate the total number of alternate futures			
		7. Perform a pairwise comparison of all alternate futures within each scenario to establish their relative probabilities			
		8. Rank the alternate futures for each scenario from highest relative probability to lowest relative probability			
		9. For each alternative future, analyze the scenario in terms of its consequences for the intelligence question			

Table 2.1—Continued

General Actions	Classic I&W Frameworks			Cyber I&W Frameworks	
	Defense Warning Network Handbook	LAMP/Lockwood	The Seven Phases of the Intelligence-Warning Process	Robinson et al. (adopted from LAMP)	INSA
Develop indicators	3. Identify conditions, drivers, and indicators; determine warning threshold	10. Determine focal events that must happen to realize each future 11. Develop indicators for each focal event 12. State the potential of a given alternate future to transpose into another alternate future	1. Identify key elements of information required to forecast a topic 2. Publish an intelligence collection plan 3. Consolidate information 4. Sort information 5. Draw conclusions	6. Create focal event indicators: An adversary prepares for hostilities	5. Collect intelligence on indicators and adversary plans and intentions.
Track indicators			6. Focus collectors on intelligence gaps to refine/update conclusions	7. Collect and monitor through indicators: assess emerging trends 8. Discern the probable scenario that is trending	
Act on indicators	4. Explore opportunities to influence or mitigate the threat 5. Communicate warning		7. Communicate conclusions/give warning	9. Readjust for new manifestations of the scenario 10. Deception in indicators 11. Mental model avoidance: Is it expectation or actuality, theory, or current developments? 12. Strategic options analyzed against viable scenarios	6. Plan and exercise countermeasures to likely adversary actions 7. Execute proactive measures to counter anticipated attack vectors

SOURCES: Adapted from Lilly et al., 2019; Khalsa, 2009.

CHAPTER THREE

RAND's Scalable Warning and Resilience Model at a Glance

In this chapter, we begin by identifying the principal components and areas for improvement of traditional I&W frameworks and similar frameworks applied to cyberspace to inform the design of a practical model. The goal of this project is to develop a model that both facilitates the warning and early detection of cyber incidents and increases resilience. Considering the mutually reinforcing role of I&W and resilience, a model integrating these two components could bring substantial improvements to existing cyber defense approaches. This chapter traces the contours of a high-level practical SWARM that integrates these two critical components and demonstrates how they can support each other in improving cyber defense. Figure 3.1 shows the key components of the model and links each of them to the components of the frameworks reviewed in the previous chapter.

Figure 3.1
SWARM and Links to Reviewed I&W Frameworks

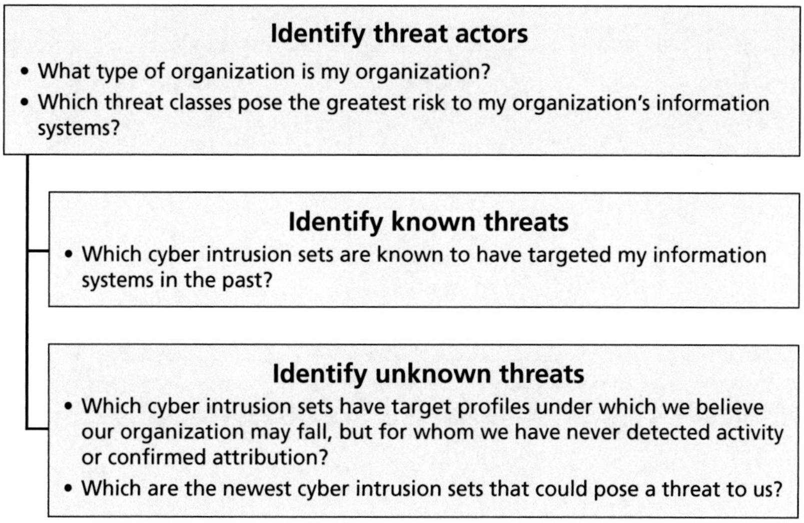

Identify threat actors
- What type of organization is my organization?
- Which threat classes pose the greatest risk to my organization's information systems?

Identify known threats
- Which cyber intrusion sets are known to have targeted my information systems in the past?

Identify unknown threats
- Which cyber intrusion sets have target profiles under which we believe our organization may fall, but for whom we have never detected activity or confirmed attribution?
- Which are the newest cyber intrusion sets that could pose a threat to us?

SOURCE: Adapted from Lilly et al., 2019; see Table 2.1.

SWARM is meant to be accessible and adaptable to organizations at various capability maturity levels, and cyber defense teams can expand, adapt, and integrate the steps in their own information environment using their resources. This model should not be an organization's sole defense mechanism, but should be one component of an effective layered defensive cyber operations approach. Ideally, prior to the application of SWARM, organizations should have already conducted an analysis and identification of their critical infrastructure and data, as well as potential vectors of attack that an adversary may exploit to reach these assets.[1]

SWARM is meant to achieve two major objectives. First, it is designed to increase predictive power by providing advance warning for cyber incidents through early and more-comprehensive indicators, both technical and nontechnical. Second, the framework intends to enhance network resilience against targeted cyber incidents. The proposed framework consists of four steps and is threat-centric (see Figure 3.2).

In Step 1, defenders are advised to set the foundation for prioritizing their cyber threat intelligence (CTI) data collection by identifying the category that their organization falls into (more on this in the next chapter) and the main cyber threat actor(s) that pose the greatest threat to their information environment. To do so, defenders can commence by consulting a list of organizational categories, which we have developed and is available in the next chapter. Once defenders identify the most-appropriate category in which their organization belongs, they can identify the set of cyber threat actors—outlined in this first step of the SWARM model—that are most likely to

Figure 3.2
RAND's Four-Step Scalable Warning and Resilience Model

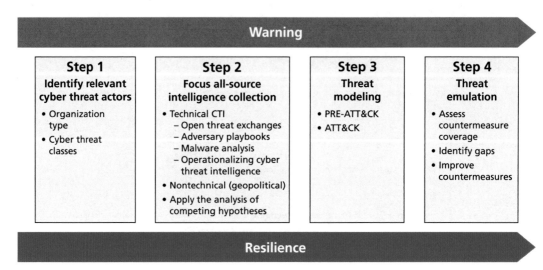

---

[1]   Deepak Bellani, *The Importance of Business Information in Cyber Threat Intelligence (CTI), the Information Required and How to Collect It*, Bethesda, Md.: SANS Institute Information Security Reading Room, 2017, pp. 3–4.

target their organization. SWARM proposes a categorization of cyber threats based on the adversaries' motivation. Organizations can juxtapose this cyber threat actors' typology with their identified organizational category and use it to focus on the category of threats that are most likely to pose a threat to their organization's operations.

In the second step of the model, defenders can use the information in Step 1—particularly the most-relevant cyber threats—to focus their all-source CTI collection by profiling the threat behavior of the most-likely adversaries of the organization. In this step, defenders can collect information on indicators that correlate with the behavior of these specific threat actors. Defenders are advised to focus on both technical and geopolitical indicators that can assist with anticipating cyber incidents. An emerging body of literature on predictive analytics, which we elaborate on in subsequent chapters, provides a promising basis for the design of a set of indicators that can be correlated with the playbooks of particular cyber threat actors and can contribute to the implementation of more-proactive cyber defense strategies.

In Step 3 of the process, defenders can dive deeper into the technical profiling of potential adversaries by applying a threat modeling framework, such as MITRE's ATT&CK and PRE-ATT&CK tools, to help identify preventions and detections already in place for the adversaries targeting their organization. If a defender is unsure whether a particular technique would be prevented or detected, Step 3 helps identify and prioritize adversarial techniques to emulate against the defender's environment (test within it) in Step 4.

As a final step, defenders are advised to conduct red or purple team activities to test relevant adversary TTPs and playbooks in their environment. This step is recommended to ensure that defenses are up-to-date through testing the tools that defenders have introduced into their systems and ensure that they have robust defenses against the threat actors that are most likely to target them.

The following chapters of this report elaborate on each of these four steps and provide a structural blueprint that different organizations can adapt to their environments to better prioritize the threats to defend against and properly allocate their defense resources.

CHAPTER FOUR

SWARM Step One: Identify Relevant Cyber Adversaries

Considering the information overflow created by various threat data feeds, network defenders' abilities to avoid alert fatigue and protect their networks can be improved by establishing a process for prioritizing the most-relevant threats to their particular organization. Different types of organizations are likely to face different cyber threat classes. Therefore, SWARM's first step is to establish a general typology of organizational sectors. This typology sets the foundation for the identification, classification, and prioritization of threat classes that are most likely to target each organizational sector. To identify the main threat actors to their information environment, defenders are advised to apply a set of standard core questions referred to as priority intelligence requirements (PIRs).

Identify Organization Type to Be Protected

Establishing a definition and conceptual categorization of organizations is the first step to prioritizing cyber threats to an organization's information environment. *Organizations* are functional administrative structures that represent systems "of consciously coordinated activities or forces."[1] We focus on organizations here, but the model could be expanded to address sectors or other communities that are likely to be targeted for similar reasons. Public administration scholars have offered various typologies to separate organizations into groups. These typologies can be based on different organizational missions or functions or the ability to observe *outputs* (the work of the organization or agency) and *outcomes* (the results of the agency's work).[2] For the purposes of this report, an appropriate conceptual foundation for dividing organizations into categories is DHS's 16 critical infrastructure sectors, which are mainly applied to a

---

[1] The public administration scholar James Q. Wilson defined *organization* as Chester Barnard had described it: as a "system of consciously coordinated activities or forces of two or more persons" (James Q. Wilson, *Bureaucracy*, new edition, New York: Basic Books, 2000, pp. 24–27; *Merriam-Webster*, "Organization," webpage, last updated September 9, 2020).

[2] Wilson, 2000, pp. 159–169.

U.S. context but can be adapted to other countries as well. These sectors are identified as vital because their "incapacitation or destruction would have a debilitating effect on security, national economic security, national public health or safety, or any combination thereof."[3] These sectors consist of physical or virtual administrative systems, networks, and assets.[4]

The concept of critical infrastructure and, as a consequence, the number of sectors that DHS defines as critical, has evolved since the first formal classification of critical infrastructure in the late 1990s.[5] In the 1980s, critical infrastructure primarily encompassed aging public physical networks, such as highways, water supply facilities, bridges, and airports. In the 1990s, the concept acquired a national security dimension preceded by an increased trend of international terrorism. After the coordinated terrorist attacks on September 11, 2001, the number of critical infrastructure sectors as described in the National Infrastructure Protection Plan further increased.[6] Although, in the 1980s, the critical sectors were focused on infrastructure adequacy, in the 1990s, federal agencies were concerned with infrastructure protection.[7] In the past few years, legislative activity has emphasized cybersecurity of critical infrastructure.[8] In 2017, as a result of growing threats to the U.S. election process, election infrastructure was designated as a subsector of the existing government facilities critical infrastructure sector.[9]

As of this report's writing, DHS defines 16 critical infrastructure sectors that this research will use to categorize threat actors classes that are likely to target each sector. These sectors are chemical; commercial facilities; communications; critical manufacturing; dams; defense industrial base; emergency services; energy; financial services; food and agriculture; government facilities; health care and public health; information technology; nuclear reactors, materials, and waste; transportation systems; and water and wastewater systems.[10]

---

[3] DHS Cybersecurity and Infrastructure Security Agency (CISA), "Critical Infrastructure Sectors," webpage, last revised March 24, 2020.

[4] DHS CISA, 2020.

[5] See President's Commission on Critical Infrastructure Protection, *Critical Foundations: Protecting America's Critical Infrastructures*, Washington, D.C., October 13, 1997.

[6] O'Rourke, 2007, p. 22.

[7] John Moteff and Paul Parfomak, *Critical Infrastructure and Key Assets: Definition and Identification*, Washington, D.C.: Congressional Research Service, RL32631, October 1, 2004.

[8] John D. Moteff, *Critical Infrastructures: Background, Policy, and Implementation*, Washington, D.C.: Congressional Research Service, RL32631, June 10, 2015.

[9] DHS, "Statement by Secretary Jeh Johnson on the Designation of Election Infrastructure as a Critical Infrastructure Subsector," Office of the Press Secretary, January 6, 2017.

[10] DHS CISA, 2020. More recently, DHS has also defined a set of 55 national critical functions to facilitate managing risk across sectors.

To ensure a more-comprehensive organizational categorization, this report supplements the 16 DHS-designated sectors with four more categories derived from an empirically driven review of reports by private-sector companies and national CERTs describing the types of organizations targeted by state-sponsored cyber adversaries.[11] These four additional categories are educational institutions, think tanks and nongovernmental organizations (NGOs), dissident groups, and intergovernmental and international organizations.

Altogether, these 20 organizational sectors serve as the starting point of SWARM and are the basis for a general categorization and identification of the classes of cyber threat actors likely to target organizations that fall under each category (see Table 4.1 toward the end of this chapter).

Categorizing Cyber Threat Classes

Cyber threat actors can be categorized on the basis of various characteristics, including level of capabilities and expertise of the cyber actors and motivations or objectives behind perpetrating a particular cyber incident. Cyber threat actor taxonomies often combine several of these characteristics to create discrete and measurable categories of threat classes.

An example of a cyber threat taxonomy based on the level of capabilities and skills that the cyber adversaries can employ is the six-tier taxonomy that was developed by the Defense Science Board. The taxonomy contains six tiers of cyber threat classes based on increasing sophistication, starting from classes that use known exploits and malicious code developed by others (least sophisticated tier) to classes that yield substantial resources and can create vulnerabilities that the adversary can exploit for military, political, economic, or other objectives (the most sophisticated tier).[12]

A more elaborate threat-based taxonomy—that groups cyber classes into distinctly different levels based on capabilities, skills, and commitment to achieving the desired objective—has been proposed by the Sandia National Laboratory. The Sandia framework aims to reduce the complexity of threat actor analysis by focusing on objectively measurable attributes. These attributes are the threat's willingness to pursue an objective, characterized by the threat's intensity and stealth and the amount of time that the threat can dedicate to planning and deploying methods to achieve its objec-

---

[11] Electronic Transactions Development Agency, *Threat Group Cards: A Threat Actor Encyclopedia*, Bangkok, June 19, 2019.

[12] DoD, Defense Science Board, *Task Force Report: Resilient Military Systems and the Advanced Cyber Threat*, Washington, D.C.: Office of the Under Secretary of Defense for Acquisition, Technology and Logistics, January 2013, p. 21.

tive, as well as resources, including people, knowledge, and access.[13] These characteristics inform the eight levels of the Sandia cyber threat matrix, in which level one is the most capable threat class and level eight is the least capable one.

Cyber threat classes can also be grouped based on the objective of the cyber incident. Such typologies categorizing cyber threat actors based on their objectives include the well-known *CIA triad* that stands for *confidentiality* (attacks compromising the privacy of data), *integrity* (attacks compromising the veracity of data), and *availability* (attacks affecting access to data).[14]

Some frameworks, such as MITRE's Cyber Prep methodology, employ a slightly different combination of these characteristics to determine threat classes. For example, the level of adversary capability, which includes skills, knowledge, opportunity, and resources; adversary objective, which includes both the objectives and the outcomes an adversary strives to avoid; and targeting, defined as the level of specificity and persistence with which an adversary pursues its target. Based on these characteristics, an organization can face five threat levels that range from unsophisticated to advanced.[15]

Cyber threat actors can also be classified primarily on the basis of their motivations. The cyber threat intelligence company Recorded Future, for example, defines four main categories of cyber threat actors: cyber criminals, hacktivists, state-sponsored attackers, and insider threats. Cyber criminals range from unorganized and under-resourced individuals to highly organized and capable groups whose primary goal is the acquisition of profit. Hacktivists also possess a variety of capabilities and are motivated by the desire to undermine an organization's or an individual's reputation or to destabilize operations. State-sponsored groups can be highly organized and well-resourced groups interested in obtaining sustained access to an organization's networks. Insider threats are intentionally or unwittingly malicious actors that vandalize, steal, or divulge proprietary assets.[16] This typology presents a clear differentiation among cyber threat classes but the categories it offers should not be considered as mutually exclusive. Certain threat actors, for example, can have both sponsorship or support from a state while also conducting operations for financial gains that would assign them to the category of cyber criminals, such as North Korea's "Lazarus Group" intrusion set.[17]

---

[13] David P. Duggan, Sherry R. Thomas, Cynthia K. K. Veitch, and Laura Woodard, *Categorizing Threat: Building and Using a Generic Threat Matrix*, Albuquerque, N.M.: Sandia National Laboratory, SAND2007-5791, September 2007, pp 19–23.

[14] InfoSec Institute, "CIA Triad," webpage, February 7, 2018.

[15] Deborah J. Bodeau, Jennifer Fabius-Greene, and Richard D. Graubart, *How Do You Assess Your Organization's Cyber Threat Level?* Bedford, Mass.: MITRE Corporation, August 2010, pp. 1–3.

[16] "Proactive Defense: Understanding the 4 Main Threat Actor Types," Recorded Future, blog post, August 23, 2016.

[17] Electronic Transactions Development Agency, 2019.

Recorded Future's motivation-based categorization of cyber threat actor classes is the most appropriate for the purposes of this analysis. Recorded Future provides a straightforward grouping of cyber threats that can be applied to the 20 organizational categories discussed previously.

Identifying the Main Cyber Threat Classes Targeting an Organization

The synthesis of organizational types culminates in the final stage of SWARM's Step 1—the identification and prioritization of threat actors that are most likely to target an organization. After identifying the type of their organization based on the categories discussed earlier in this chapter, defenders can apply the set of questions similar to the ones outlined in Figure 4.1. These questions are derived from PIRs commonly used in the USIC and military to focus data collection on the most-critical elements of information needed for decisionmaking.[18] PIRs are a concise set of questions that an organization could pose to itself to identify the main cyber actors that present the greatest threat to the organization. PIRs can facilitate the identification of known (those that have breached the networks of the organization in the past) as well as unknown (those that are familiar to the cybersecurity community but have

Figure 4.1
Example Questions to Facilitate the Identification of Most Likely Cyber Threats

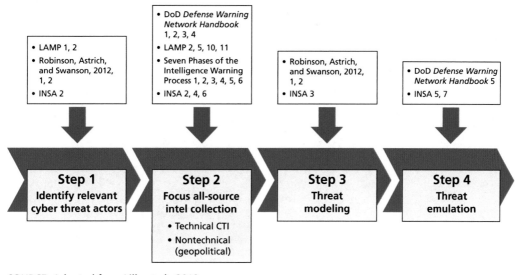

SOURCE: Adapted from Lilly et al., 2019.

---

[18] Joint Publication 3-0, *Joint Operations*, incorporating Change 1, Washington, D.C.: Joint Chiefs of Staff, October 22, 2018.

Table 4.1
Critical Infrastructure Sectors, Most Likely Cyber Threats Classes per Sector, and Examples of Intrusion Sets in Each Class

Critical Infrastructure Sector	Relevant Types of Threat Classes	Intrusion Sets
Chemical	State-sponsored	• APT 1 (FireEye), Comment Crew (Symantec) • APT 33 (FireEye), Elfin (Symantec) • TG-2889 (SecureWorks), Cutting Kitten (CrowdStrike) • APT 15 (FireEye), Vixen Panda (CrowdStrike) • Lead (Microsoft)
Commercial facilities	State-sponsored	• APT 30 (FireEye), Override Panda (CrowdStrike) • APT 33 (FireEye), Elfin (Symantec) • APT 28 (FireEye), Fancy Bear (CrowdStrike)
	Cybercrime	• OurMine (hacking group self-designation)
Communications	State-sponsored	• APT 3 (FireEye), Gothic Panda (CrowdStrike) • APT 12 (FireEye), Numbered Panda (CrowdStrike) • APT 18 (FireEye), Dynamite Panda (CrowdStrike) • APT 39 (FireEye), Chafer (Symantec)
	Cybercrime	• FIN 7 (FireEye)
Critical manufacturing	State-sponsored	• Winnti Group (Kaspersky) • Lead (Microsoft)
Dams	State-sponsored	• TG-2889 (SecureWorks), Cutting Kitten (CrowdStrike)
Defense industrial base	State-sponsored	• DarkHotel (Kaspersky) • TG-2889 (SecureWorks), Cutting Kitten (CrowdStrike)
Dissident groups	State-sponsored	• Stealth Falcon (Citizen Lab), FruityArmor (Kaspersky) • APT 32 (Mandiant), OceanLotus (SkyEye Labs) • Flying Kitten (CrowdStrike), Ajax Security Team (FireEye)
Educational institutions	State-sponsored	• APT 18 (FireEye), Dynamite Panda (CrowdStrike) • APT 19 (FireEye), C0d0so (CrowdStrike)
Emergency services	State-sponsored	• APT 28 (FireEye) (aka "Cyber Caliphate")
Energy	State-sponsored	• APT 29 (FireEye), Cozy Bear (CrowdStrike) • APT 19 (FireEye), C0d0so (CrowdStrike) • APT 33 (FireEye), Elfin (Symantec) • TG-2889 (SecureWorks), Cutting Kitten (CrowdStrike) • DarkHotel (Kaspersky) • APT 35 (FireEye), Charming Kitten (CrowdStrike) • Lotus Panda (CrowdStrike)
	Cybercrime	• FIN 7 (FireEye) • Mummy Spider (CrowdStrike)

Table 4.1—Continued

Critical Infrastructure Sector	Relevant Types of Threat Classes	Intrusion Sets
Financial services	State-sponsored	• APT 19 (FireEye), C0d0so (CrowdStrike) • Bling Eagle (360) • TG-2889 (SecureWorks), Cutting Kitten (CrowdStrike)
	Cyber crime	• Buhtrap (Group-IB) • Carbanak (Kaspersky), Anunak (Group-IB) • Cobalt Group (Group-IB) • Corkow (Group-IB), Metel (Kaspersky) • FIN 4 (FireEye), Wolf Spider (CrowdStrike)
Food and agriculture	State-sponsored	• APT 1 (FireEye), Comment Crew (Symantec)
Government facilities	State-sponsored	• APT 3 (FireEye), Gothic Panda (CrowdStrike) • APT 14 (FireEye), Anchor Panda (CrowdStrike) • APT 12 (FireEye), Numbered Panda (CrowdStrike) • APT 29 (FireEye), Cozy Bear (CrowdStrike)
Healthcare and public health	State-sponsored	• APT 18 (FireEye), Dynamite Panda (CrowdStrike) • APT 1 (FireEye), Comment Crew (Symantec) • TG-2889 (SecureWorks), Cutting Kitten (CrowdStrike) • Dark Caracal (Lookout) • APT 37 (FireEye), Reaper (FireEye)
	Cyber criminals	• Gold Lowell (SecureWorks), Boss Spider (CrowdStrike) • Mummy Spider (CrowdStrike), TA542 (Proofpoint)
Information technology	State-sponsored	• APT 3 (FireEye), Gothic Panda (CrowdStrike); • APT 17 (FireEye), Deputy Dog (iDefense)
Intergovernmental and international organizations	State-sponsored	• GhostNet (Information Warfare Monitor), Snooping Dragon (UCAM) • APT 28 (FireEye), Fancy Bear (CrowdStrike)
Nuclear reactors, materials, and waste	State-sponsored	• Berserk Bear, Dragonfly 2.0 (Symantec)
Transportation systems	State-sponsored	• APT 3 (FireEye), Gothic Panda (CrowdStrike) • APT 18 (FireEye), Dynamite Panda (CrowdStrike) • APT 29 (FireEye), Cozy Bear (CrowdStrike)
Water and wastewater systems	State-sponsored	• APT 14 (FireEye), Anchor Panda (CrowdStrike)
Think tanks and NGOs	State-sponsored	• APT 29 (FireEye), Cozy Bear (CrowdStrike) • APT 35 (FireEye), Charming Kitten (CrowdStrike) • Clever Kitten (CrowdStrike) • Kimsuky (Kaspersky), Velvet Chollima (CrowdStrike)

SOURCES: DHS CISA, 2020; "Proactive Defense: Understanding the 4 Main Threat Actor Types," 2016; Electronic Transactions Development Agency, 2019; Government of the United Kingdom, Foreign and Commonwealth Office, National Cyber Security Centre, and Jeremy Hunt, "UK Exposes Russian Cyber Attacks," press release, October 4, 2018; and Christian Quinn, "The Emerging Cyberthreat: Cybersecurity for Law Enforcement," *Police Chief*, December 12, 2018.
NOTES: TG = Threat Group; FIN 7 is a financially focused hacking group.

not breached the networks of the organization in the past) threats to an information environment.

When answering the questions pertaining to the identification of known and unknown threats, defenders can create advanced persistent threat (APT) lists focused on the intrusion sets most likely to target their networks. Table 4.1 highlights some of the main cyber threat classes and several of the key cyber threat actors within each class that are likely to target organizations pertaining to a particular organization type.

SWARM's Step 1 requires defenders to identify their type of organization and the threat classes and intrusion sets that are most likely to target their organization's information environment. In the subsequent steps of this model, defenders will use this knowledge to gather further information about the TTPs of their most likely adversaries and apply predictive analytics and other methods that will contribute to the early detection and warning of these adversaries' actions, curtailing response time and time to remediation while enhancing resilience.[19]

---

[19] Time to remediation is the time necessary to eject the cyber intrusion set from an organization's networks and clean it. The average time to remediation for the most-efficient organizations is estimated to be about 60 minutes. Michael Busselen, "CrowdStrike CTO Explains 'Breakout Time'—A Critical Metric in Stopping Breaches VIDEO]," *CrowdStrike*, June 6, 2018.

CHAPTER FIVE

SWARM Step Two: Focus All-Source Intelligence Collection

After identifying the key cyber threat actors likely to target their organization in Step 1, network defenders can proceed with collecting all-source intelligence about these threat actors. U.S. Army Field Manual 2-0, Chapter 5-1, defines all-source intelligence as "the intelligence products, organizations, and activities that incorporate all sources of information and intelligence, including open-source information, in the production of intelligence."[1] This includes technical tactical- and operational-level CTI and strategic-level open-source intelligence (OSINT). Calculated employment of highlighted information sources within each discipline and threat intelligence level can enable early detection of potential adversarial activities against the information environment of one's organization.[2] Presuming no access to classified resources, Step 2 of SWARM centers on this combination of focused technical CTI collection and operationalization and nontechnical (geopolitical) OSINT collection and analysis about the cyber threat actors' respective nation-state geopolitics, relative to the defenders' own nation and organization type. It should be noted that these two components of Step 2 can be performed concurrently by different teams or individuals.

Novel components of Step 2 of SWARM include emphasizing several unique CTI operationalization actions that, most importantly, can boost early warning, and incorporating a methodology applied to open-source information that enables some level of predictive analysis using high-level nontechnical indicators.

For the OSINT-based nontechnical component, we present a method for identifying relevant high-level semantic indicators derived from OSINT data collection domains and methods that can be applied to the data in an attempt to generate early warning intelligence of an imminent incident. This process results in a probabilistic prediction dimension added to the analysis that may facilitate the anticipation of the time frames during which threat actors may target the defenders' information environments. Deriving these predictive indicators assists defenders in answering the question: *When will a particular adversary attack next?* This can allow time for defenders

---

[1] U.S. Department of the Army, *Intelligence*, Field Manual No. 2-0, Washington, D.C., May 2004, p. 1–30.

[2] INSA, Cyber Intelligence Task Force, *Operational Levels of Cyber Intelligence*, Arlington, Va., September 2013.

to prioritize the emulation of threat-modeled adversarial techniques, the application of countermeasures, infrastructure hardening, and the allocation of resources, resulting in enhanced resilience.[3]

Predicting the future with absolute certainty is virtually impossible, but the model illustrates that, in the cases of some adversaries in the espionage threat class, it may be possible to achieve early warning of an impending attack with a qualitative probability rating within a certain approaching window of time. SWARM's Step 2 outlines a framework for encapsulating analytic techniques applied to CTI operationalization that focuses on early warning of imminent adversary activity and increases resilience, as well as environmental triggers that can affect or serve as barometers for an adversary's decision and timing to initiate a cyber incident.

This Step 2 framework, which includes both data collection domains and methods that can be applied to the data, can then be used to derive predictive indicators and IOCs to assist defenders in enhancing the resilience and predictive capacity of their analysis. The effectiveness of this approach will be enhanced through information-sharing and regular revision of observed geopolitical dynamics, developed methodological techniques, and data collection domains.

Technical CTI Collection

Cyber threat intelligence consists of collecting and analyzing information from a rich array of sources and is designed on the basis of analytical tradecraft that has been refined by government and military agencies over several decades.[4] In addition to providing an overwhelming amount of information, CTI collection customarily relies on post-breach IOC formulation and sharing. CTI focuses on the collection and analysis of technical data about intrusion sets, and, for the purposes of this report, it is considered only one element of a cyber I&W framework. The identification of threat actors targeting or most likely to target an organization's information environment can assist with the prioritization of CTI that focuses on IOCs associated with the identified relevant cyber threat actors.

---

[3]   A *countermeasure* is an action taken to oppose, neutralize, or retaliate against some other action. In this context, it is taken to include defensive actions that amount to prevention, detection, or deception. Countermeasures can also be thought of as actions that fall into the detect, deny, disrupt, degrade, deceive, or destroy categories in Eric M. Hutchins, Michael J. Cloppert, and Rohan M. Amin's Courses of Action Matrix in "Intelligence-Driven Computer Network Defense Informed by Analysis of Adversary Campaigns and Intrusion Kill Chains," Bethesda, Md.: Lockheed Martin Corporation, 2011, adapted from the 2006 version of Joint Publication 3-13, *Information Operations*, Washington, D.C.: Joint Chiefs of Staff, Incorporating Change 1, November 20, 2014. Patching, antivirus, honeypots, detection signatures (both IOA and IOC), and the like are considered countermeasures of various types.

[4]   Pace, 2018, p. 2; Zane Pokorny, "What Is Threat Intelligence? Definition and Examples," blog post, Recorded Future, April 30, 2019.

Moreover, an all-source approach that encompasses not only technical but also nontechnical or geopolitical indicators can improve the early detection and warning of cyber incidents by allowing defenders to scan holistically the entire operational space for potential behavioral and environmental triggers that can signal an adversary's intent to initiate a cyber incident.

Cyber threat intelligence—and the standard practice of collecting and operationalizing it—is a discipline that has been extensively written about. The scope of this section is not to reiterate everything about CTI, how to collect it, why to do it, or how not to do it, but rather to highlight key components of it in relation to SWARM, emphasizing how these components facilitate the process of early warning and resilience. The processes of data collection and operationalization are briefly described in this section, and the defense teams of each organization are advised to tailor them to their information environment.

Once defenders have identified which adversaries are targeting their organization in Step 1, they can use closed-source or open-source unclassified information to focus CTI collection on the threats that pose the greatest risks to their organization. Defenders should focus CTI collection on those adversaries and exploit the multiple aliases each adversary likely has to ensure a more-complete collection of IOCs and TTPs related to the particular threat. This will include atomic IOCs,[5] which are considered low-level semantic indicators, as well as TTPs. Additionally, we highlight aspects of CTI operationalization that pertain to deriving early warning intelligence. This chapter proceeds with an elaboration on several essential foundational aspects of data collection that are likely to strengthen the defense capabilities of information systems.

Open Threat Exchanges

Alien Vault's Open Threat Exchange (OTX) or ThreatConnect's TC Open are examples of useful repositories from which to begin collecting IOCs related specifically to the adversaries targeting an organization.[6]

Adversary Playbooks

Playbooks published by security researchers often include IOCs associated with their modeled intrusion sets. Such companies as Palo Alto Networks and such nonprofit organizations as the Cyber Threat Alliance have developed and published playbooks. These playbooks represent an "attempt to capture the full collection of tools, tactics,

---

[5] *Atomic IOCs* are low-level IOCs that cannot be broken down further, such as hashes, Internet Protocol (IP) addresses, or uniform resource locators (URLs).

[6] AT&T Cybersecurity, "AT&T Alien Labs Open Threat Exchange," webpage, undated.

techniques, and procedures that adversaries use to achieve their goals, arrayed in a logical sequence using the Lockheed Martin Cyber Intrusion Kill Chain."[7]

Malware Analysis

Whether conducting manual code-level static reverse engineering, manual code-level interactive behavioral analysis, or automated dynamic analysis in a sandbox, malware samples—especially recent ones—from an adversary's malware arsenal can be analyzed both to derive atomic IOCs and to uncover the capabilities of the tools being used against their organization or similar organization types. Malware samples from an adversary can be obtained from the defenders' own incident archive data or, if the defenders have hashes, possibly from such sites as VirusShare.com. Additionally, unique strings or byte sequences extracted from malware can be used to create Yara signatures (a tool that can be used to classify malware samples), which can be used for VirusTotal Hunting (which is described in the "VirusTotal Hunting" section of this chapter).[8] Finally, the tool capabilities uncovered by understanding the malware with which an organization or their organization type is targeted can subsequently be threat-modeled in SWARM Step 3.

Operationalize the CTI in the Information Environment of the Organization

Defenders are advised to operationalize the CTI in their organization's environment to harden its infrastructure, increase resilience, and gain cyber situational awareness about potential interactions in the past or present with the identified actual or potential adversaries. Additionally, Passive Domain Name System (pDNS or passive DNS) analysis can sometimes feed early warning based on adversary infrastructure being staged prior to an attack.

Atomic IOCs

Defenders can consider incorporating the atomic IOCs into their network defensive posture, provided that they trust or have validated them.[9] Examples of incorporating atomic IOCs into a network defensive posture to harden an information environment include blocking hostile IP addresses inbound and outbound in an enterprise firewall, blocking C2 domains in one's proxy, globally banning hashes of malicious files in an application "allow listing" tool, redirecting malicious email sender addresses to a mailbox that a cyber defense team controls, and using an endpoint detection and response (EDR) tool or the open-source Kansa PowerShell–based incident response (IR) frame-

---

[7]   Unit 42, "Playbook Viewer," webpage, GitHub, undated; and Cyber Threat Alliance, "Playbooks," webpage, undated.

[8]   "Yara in a Nutshell," webpage, GitHub, undated.

[9]   Joe Slowik, "Threat Analytics and Activity Groups," blog post, *Dragos*, February 26, 2018.

work to hunt for unique registry key IOCs.[10] It is important to also ensure that defenders create content in their security information and event manager (SIEM) so that it will distribute an alert on any future interactions between their network or endpoints and any of the IOCs. Conducting queries in the SIEM for any past or present activity with the IOCs also provides visibility into potential interactions with an adversary or its infrastructure on a timeline that has the potential to support predictive analysis for probable future adversarial activities against the information environment.

Passive DNS Analysis

The power of passive DNS analysis cannot be understated. A valuable tool for this is RiskIQ's platform, PassiveTotal, which has a free Community Edition with numerous features and integrations, enabling easy querying and pivoting from known IOCs, such as IPs and domains, to previously unknown infrastructure.[11] This is done by leveraging PassiveTotal's collected associations between passively collected DNS resolution histories by IPs and their historical associations with malware hashes, websites' domain name information that can be obtained from WHOis.net, X.509 Secure Sockets Layer (SSL) certificate hashes, serial numbers and metadata, and other features. There are also many other integrations in PassiveTotal that can provide discovery of adversary infrastructure or artifacts that may be connected and previously unknown to the defenders (see Figure 5.1).[12] These connections can also be graphed in the software suite Maltego for visualization.[13]

Related to early warning, domain registration and resolution date and time stamps can help paint a timeline of C2 infrastructure staging, which corresponds to MITRE's PRE-ATT&CK "Establish and Maintain Infrastructure" tactic (TA0022) and its Buy Domain Name technique (T1328).[14] Occasionally, this analysis can result in discovering evidence that an adversary's C2 infrastructure has been registered recently but has not yet resolved to any IP addresses, signaling early warning that it has been staged for an upcoming attack or campaign. A challenge to this analysis is when an adversary compromises preexisting legitimate domains on the internet instead of registering its own. However, some of the associations possible in PassiveTotal may still prove valuable.

---

[10] "A Powershell Incident Response Framework," webpage, GitHub, last updated August 4, 2020.

[11] RiskIQ, "RiskIQ Community Edition," web tool, undated.

[12] *Passive DNS* is a system of record that stores DNS resolution data for a given location, record, and time period. This historical resolution data set allows analysts to view which domains resolved to an IP address and vice versa. This data set allows for time-based correlation based on domain or IP overlap.

[13] Maltego has a free community edition of its link-analysis/visualization product. See Maltego, "Maltego Products," webpage, undated.

[14] MITRE | ATT&CK, "Establish and Maintain Infrastructure, Techniques," webpage, last modified October 17, 2018c; and MITRE | ATT&CK, "Buy Domain Name, Procedure Examples," webpage, last modified October 17, 2018b.

Figure 5.1
RiskIQ's PassiveTotal Tool, Showing Resolution History of an APT 34 Domain

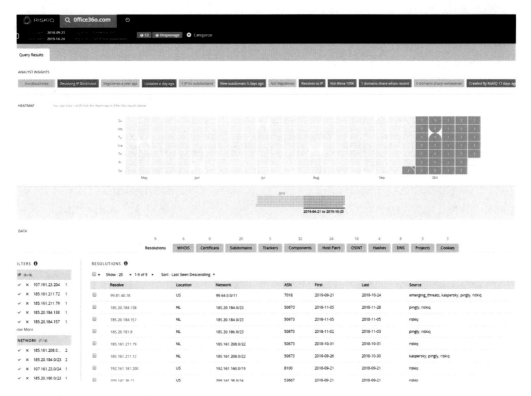

SOURCE: RiskIQ, undated. Used with permission.

VirusTotal Hunting

Another novel technique that defenders can employ to operationalize hashes from CTI is VirusTotal Hunting with VirusTotal's services Livehunt and Retrohunt. This is not a free service, but it is worth mentioning. A sample workflow is to take any malware created by an organization's adversary and generate Yara signatures, either manually or with a tool.[15] The Yara rules can then be inserted into Retrohunt to discover existing malware samples on VirusTotal that match that signature, or into Livehunt, which will detect any malware samples that are uploaded that match a Yara signature, resulting in an immediate email notification. These processes allow defenders to potentially uncover current adversary actions if they use Livehunt to test detection of its malware. This occurs in the PRE-ATT&CK time frame, specifically, by using the Test Capa-

---

[15] For an example of a tool with which Yara signatures can be generated, see "A Yara Rule Generator for Finding Related Samples and Hunting," webpage, GitHub, last updated November 6, 2018.

bilities tactic (TA0025) and Test Malware to Evade Detection technique (T1359).[16] VirusTotal thus allows defenders to use this CTI-based workflow that defenders can use to potentially detect early warning of an attack.

Information-Sharing

Information-sharing of CTI with trusted peers or partner organizations—whether the sharing is manual or automated—for collective defense and threat awareness can enhance resilience and early warning for any organization. Such a process will require vetting and trusting the indicators or automating cross-checks against known "allow lists" prior to any prevention or detection engineering but can supplement basic situational awareness and infrastructure hardening.[17]

However, some of the most valuable CTI to share is what one's cyber defense team generates based on what that team has detected on or against its network. Often, adversaries do not conduct phishing campaigns against all their targets simultaneously, and sometimes they reuse infrastructure or assets. This leaves an opportunity for information-sharing to benefit a peer organization if the adversary has not yet targeted or breached that organization. In this way, what one organization shares can serve as early warning intelligence for another.

TTPs transcend atomic IOCs, and SWARM builds on that by proceeding with threat modeling in Step 3 and then operationalizing the modeled TTPs in Step 4 by emulating them to tune network defensive posture, enhancing resilience. Next—in Step 2, which should be performed concurrently with CTI collection and operationalization if the organization possesses sufficient personnel resources—is a look at methods for analyzing OSINT to aid in prediction and early warning of cyber incidents.

Nontechnical OSINT Collection and Analysis

Novel Data Collection Domains and Methodological Trends That Can Facilitate Advance Warning of Cyber Incidents

Analysis by Robinson, Astrich, and Swanson focuses on attempting to enhance resilience and the predictive power of defense strategies by increasing warning time and incorporating nontechnical indicators in the monitoring and detection of cyber threats. An emerging trend is the use of behavioral and predictive analytics, which focuses on building a spectrum of future scenarios based on all-source intelligence, expert judgment, intuition, and viable predetermined event and wild-card scenarios. The analysis identifies the adversaries and their environmental triggers, capabilities, intentions, and

---

[16] MITRE | ATT&CK, "Test Capabilities, Techniques," webpage, October 17, 2018d; and MITRE Corporation, "ATT&CK: Test Malware to Evade Detection," webpage, October 17, 2018e.

[17] Although not widely used yet, we have chosen to use the neutral terms "allow" and "deny" list in place of previous use of the terms "whitelist" and "blacklist."

motivations while using diverse domains for data collection.[18] Some recently explored methods for analyzing and predicting threat actors' behavior include graph models, such as Bayesian networks, Markov models and attack graphs, time series, machine learning and data mining techniques, DDoS volume forecasting, and evolutionary computing.[19]

A variety of analyses propose to detect cyber incidents before the point of breach of the system or at the early stages of the breach by relying on the application of risk assessment methods, game theoretic models, or machine learning techniques, such as generic algorithms, hidden Markov models, deep neural networks, and autoregressive time series models. Some of these analyses apply these techniques to technical data, such as network traffic information. For example, Abdlhamed et al. (2016) offer a possibility of improving intrusion prediction techniques in a cloud computing environment through the use of risk assessment and predictive statistical analysis.[20] Bilge et al. (2017) illustrate how a system that analyzes binary file logs of machines can be used to predict which machines are at risk of infection months before they are infected through the application of supervised machine learning methods.[21] Divya and Muniasamy (2015) propose a framework for increasing the accuracy of detecting an intrusion and for predicting the next stages of intrusions by combining the machine learning methods of a genetic algorithm and a hidden Markov model.[22] Eder-Neuhauser et al. (2017) analyze typically used malware that is deployed against smart grids to prepare the defenses of utility companies against future malware types.[23] Abeshu and Chi-

---

[18] Robinson, Astrich, and Swanson, 2012.

[19] Martin Husák, Jana Komárková, Elias Bou-Harb, and Pavel Čeleda, "Survey of Attack Projection, Prediction, and Forecasting in Cyber Security," *IEEE Communications Surveys and Tutorials*, Vol. 21, No. 1, 2018, p. 15; Drew Robb, "Eight Top Threat Intelligence Platforms," *eSecurity Planet*, July 18, 2017; Aldo Hernandez-Suarez, Gabriel Sanchez-Perez, Karina Toscano-Medina, Victor Martinez-Hernandez, Hector Perez-Meana, Jesus Olivares-Mercado, and Victor Sanchez, "Social Sentiment Sensor in Twitter for Predicting Cyber-Attacks Using ℓ1 Regularization," *Sensors*, Vol. 18, No. 5, May 2018, p. 1380; Amy Blackshaw, "Behavior Analytics: The Key to Rapid Detection and Response?" webpage, RSA, January 22, 2016; and interview with a cybersecurity expert, December 21, 2018 (name withheld on request).

[20] Mohamed Abdlhamed, Kashif Kifayat, Qi Shi, and William Hurst, "A System for Intrusion Prediction in Cloud Computing," *Proceedings of the International Conference on Internet of Things and Cloud Computing*, New York, 2016.

[21] Leyla Bilge, Yufei Han, and Matteo Dell'Amico, "RiskTeller: Predicting the Risk of Cyber Incidents," *Proceedings of the 2017 ACM SIGSAC Conference on Computer and Communications Security*, Dallas, Tex., October 30–November 3, 2017.

[22] T. Divya and K. Muniasamy, "Real-Time Intrusion Prediction Using Hidden Markov Model with Genetic Algorithm," in L. Padma Suresh, Subhransu Sekhar Dash, and Bijaya Ketan Panigrahi, eds., *Artificial Intelligence and Evolutionary Algorithms in Engineering Systems: Proceedings of ICAEES 2014*, Vol. 1, New Delhi: Springer, India, 2015.

[23] Peter Eder-Neuhauser, Tanja Zseby, Joachim Fabini, and Gernot Vormayr, "Cyber Attack Models for Smart Grid Environments," *Sustainable Energy, Grids and Networks*, Vol. 12, December 2017.

lamkurti (2018) propose a deep learning scheme for improving the detection of cyber incidents.[24] Oehmen et al. (2016) examine anomalous events in network traffic using behavior-based statistical models that include a contextual perspective to improve the detection of cyber incidents.[25] These analyses represent only a fraction of the research conducted with large data sets and using sophisticated methods to attempt to improve primarily the early detection of cyber incidents based on cyberspace data.

Other analyses apply these novel methods to nontechnical data, such as the behavior of threat actors and their environment, which can also facilitate the identification of cyber incidents before the delivery of the weaponized payload phase of Lockheed Martin's Cyber Kill Chain. For example, Okutan et al. (2017) propose to predict cyber incidents at least one day before the point of breach (the Cyber Kill Chain's delivery of the weaponized payload phase) based on a Bayesian classifier that incorporates unconventional signals, such as attack mentions on Twitter and the number of attacks reported on the website Hackmageddon.[26] Goyal et al. (2018) describe machine learning techniques based on deep neural networks and autoregressive time series models that use

---

[24] Abebe Abeshu and Naveen Chilamkurti, "Deep Learning: The Frontier for Distributed Attack Detection in Fog-to-Things Computing," *IEEE Communications Magazine*, Vol. 56, No. 2, February 2018. In addition, Aditham and Ranganathan (2015) offer a hardware-driven model for detecting attacks on the basis of attack probability score; Fava et al. (2008) use Markov models to analyze the sequential properties of attack paths, which could allow prediction of further actions in ongoing attacks; Fuertes et al. (2017) apply a business intelligence model focused on malware vulnerability analysis to produce early intrusion warnings. See also Halawa et al. (2017), Hernandez et al. (2013), Lakhno et al. (2016), and Mathew et al. (2005). Santosh Aditham and Nagarajan Ranganathan, "A Novel Framework for Mitigating Insider Attacks in Big Data Systems," *Proceedings of the 2015 IEEE International Conference on Big Data (Big Data)*, 2015; Daniel S. Fava, Stephen R. Byers, and Shanchieh Jay Yang, "Projecting Cyberattacks Through Variable-Length Markov Models," *IEEE Transactions on Information Forensics and Security*, Vol. 3, No. 3, September 2008; Walter Fuertes, Francisco Reyes, Paúl Valladares, Freddy Tapia, Theofilos Toulkeridis, and Ernesto Pérez, "An Integral Model to Provide Reactive and Proactive Services in an Academic CSIRT Based on Business Intelligence," *Systems*, Vol. 5, No. 4, 2017, p. 20; Hassan Halawa, Matei Ripeanu, Konstantin Beznosov, Baris Coskun, and Meizhu Liu, "An Early Warning System for Suspicious Accounts," *Proceedings of the 10th ACM Workshop on Artificial Intelligence and Security*, Dallas, Tex., November 2017; Jarilyn M. Hernández, Line Pouchard, Jeffrey McDonald, and Stacy Prowell, "Developing a Power Measurement Framework for Cyber Defense," *Proceedings of the Eighth Annual Cyber Security and Information Intelligence Research Workshop*, Oak Ridge, Tenn., January 2013; Valeriy Lakhno, Svitlana Kazmirchuk, Yulia Kovalenko, Larisa Myrutenko, and Tetyana Okhrimenko, "Design of Adaptive System of Detection of Cyber-Attacks, Based on the Model of Logical Procedures and the Coverage Matrices of Features," *East European Journal of Advanced Technology*, Vol. 3, No. 9, June 2016; and Sunu Mathew, Daniel Britt, Richard Giomundo, and Shambhu Upadhyaya, "Real-Time Multistage Attack Awareness Through Enhanced Intrusion Alert Clustering," *MILCOM 2005 - 2005 IEEE Military Communications Conference*, Atlantic City, N.J., October 17–20, 2005.

[25] Christopher S. Oehmen, Paul J. Bruillard, Brett D. Matzke, Aaron R. Phillips, Keith T. Star, Jeffrey L. Jensen, Doug Nordwall, Seth Thompson, and Elena S. Peterson, "LINEBACKER: LINE-Speed Bio-Inspired Analysis and Characterization for Event Recognition," *IEEE Security and Privacy Workshops (SPW)*, San Jose, Calif., May 22–26, 2016.

[26] Ahmet Okutan, Shanchieh Jay Yang, and Katie McConky, "Predicting Cyber Attacks with Bayesian Networks Using Unconventional Signals," *CISRC '17 Proceedings of the 12th Annual Conference on Cyber and Information Security Research*, No. 13, April 2017.

external signals from publicly available internet sources to forecast cyber incidents. The researchers demonstrate that different data sources should be monitored in anticipation of different cyber incidents against different targets and that the most reliable signals are unique to different organizations, thus concluding that the model has to be applied and customized to the unique environment of each organization.[27] Maimon et al. (2017) argue that, in addition to using information about attackers' resources, access to the target, skills, knowledge, and motivation to offend, cyber defense teams should draw knowledge from criminological models and use analyses about the attacker's personality traits and demographics, as well as circumstances that influence the offenders' decision to instigate a cyber incident.[28] Sharma et al. (2013) focus on human intentions and argue that social, political, economic, and cultural factors in the physical space are often the causes for socially related cyber incidents. On the basis of this assumption, the authors develop a threat-based attack model to monitor such events and predict cyber incidents before the delivery of a weaponized payload.[29] Waters et al. (2012) use ethnographic models and regression analysis to analyze the links between gross domestic product, corruption, and criminal activity in cyberspace. The results suggest that a skilled workforce that operates in a country with high levels of perceived corruption is associated with higher levels of cybercrime. These findings can be used to improve the profile of certain cyber criminals and better predict cyber incidents initiated by these particular cyber threat actors.[30] Dalton et al. (2017) propose to improve cyber incident forecasting through cognitive augmentation and information foraging using publicly available data sources, such as the data available on the website Hackmageddon, that have not been typically used for cyber incident prediction.[31]

---

[27] Palash Goyal, Tozammel Hossain, Ashok Deb, Nazgol Tavabi, Nathan Barley, Andrés Abeliuk, Emilio Ferrara, and Kristina Lerman, "Discovering Signals from Web Sources to Predict Cyber Attacks," *arXiv*, preprint, August 2018.

[28] David Maimon, Olga Babko-Malaya, Rebecca Cathey, and Steve Hinton, "Re-Thinking Online Offenders' SKRAM: Individual Traits and Situational Motivations as Additional Risk Factors for Predicting Cyber Attacks," *IEEE 15th International Conference on Dependable, Autonomic and Secure Computing, 15th International Conference on Pervasive Intelligence and Computing, 3rd International Conference on Big Data Intelligence and Computing, and Cyber Science and Technology Congress (DASC/PiCom/DataCom/CyberSciTech)*, Orlando, Fla., November 6–10, 2017.

[29] Anup Sharma, Robin Gandhi, Qiuming Zhu, William Mahoney, and William Sousan, "A Social Dimensional Cyber Threat Model with Formal Concept Analysis and Fact-Proposition Inference," University of Nebraska Omaha, Computer Science Faculty Publications, No. 24, December 2013.

[30] Paul A. Watters, Stephen McCombie, Robert Layton, and Josef Pieprzyk, "Characterising and Predicting Cyberattacks Using the Cyber Attacker Model Profile (CAMP)," *Journal of Money Laundering Control*, Vol. 15, No. 4, 2012.

[31] Adam Dalton, Bonnie Dorr, Leon Liand, and Kristy Hollingshead, "Improving Cyber-Attack Predictions Through Information Foraging," *IEEE International Conference on Big Data (Big Data)*, Boston, Mass., December 11–14, 2017.

These studies offer a variety of sophisticated solutions, but undertaking what is described in them requires a specific skill set and time to perform, update, and maintain the series of analyses, which are usually beyond the mission scope of cyber defense teams. The general effectiveness of these techniques and models highly depends on the type of organization, its information environment, available data on cyber adversaries, available resources, and analytical capacity. To use this research in their daily defense operations, cyber defense teams would need to work with a multidisciplinary staff, including statisticians, econometricians, computer scientists, and even criminologists and ethnographers, who can perform the analyses and translate them into actionable deliverables for a cyber defense team. Some of the proposed approaches would also require developing and implementing new tools that can enable the continuous collation and analysis of information from a variety of sources in real time; however, the insights these approaches could yield would likely be very beneficial to the cyber defense community. Furthermore, the analyses presented in these reports typically do not provide specific guidelines on how to integrate the models proposed in these reports into an existing cyber defense architecture, and they do not link the findings to the behavior of specific cyber threat actors. Many organizations may not be able to afford the allocation of resources necessary to ensure that such skill sets, the capacity for developing additional tools, and the blueprints for the integration of the results of these studies are available within the organization.

A Simplified Methodological Framework for Analyzing Nontechnical Data to Improve Cyber Incident Forecasting

Adding a focus on the strategic, nontechnical environment can improve CTI and defenders' ability to predict cyber incidents before the point of breach by moving beyond the technical aspects of cybersecurity to attempt to understand the external factors that influence the motivations behind and the timing of hostile cyber activity. To be applicable and useful to a cyber defense team, the inclusion of such analysis should be more specific and linked to specific threat actors that are most likely to target their particular organization.

This approach is predicated on the fact that certain classes of cyber threat actors, particularly state-sponsored threat actors, conduct campaigns for strategic reasons—usually of a geopolitical nature—that extend beyond the cyber domain. In some cases, an APT may seek to exploit a major international event, such as the Olympic Games. This was the case with Russia's APTs, which tried to obfuscate an attack on the PyeongChang 2018 Winter Olympics, trying to create the impression the attack was in fact conducted by North Korean intrusion sets—a false flag operation. Experts hypothesize that the impetus behind Russia's attacks was the banning of Russia's team from competing at the Winter Olympics as a result of doping allegations.[32] Because of Russia's 2018 Winter Olympics cyber incidents, large cybersecurity firms were already focusing their efforts on ensuring their information environments. Services offered to entities involved in the 2020 Summer Olympics in Tokyo were prepared to detect and mitigate a potential cyber incident initiated by Russian APTs a year ahead of time, prior to the 2020 Olympics' cancellation as a result of the coronavirus pandemic.[33]

In other cases, analysis demonstrates that the imposition of economic sanctions may cause financial concerns for the sanctioned state and result in an expected increase of types of state-sponsored cyber incidents. In particular, the announcement of new rounds of United Nations (UN) sanctions against North Korea has been associated with an increase in profit-seeking cyber operations by North Korean threat actors, such as the Lazarus Group. Such operations include attacks on cryptocurrency exchanges in South Korea and the 2016 Society for Worldwide Interbank Financial Communications (SWIFT) banking system hack.[34] The cases of cyber incidents linked to such events as the Olympic Games and the announcement of sanctions illustrate the relevance of incorporating nontechnical information in the design of indicators that can facilitate the prediction of adversarial behavior in cyberspace.

A relatively straightforward analytical method can be applied to the collected open-source nontechnical data to ensure their exhaustive and systematic analysis; that is the *analysis of competing hypotheses* (ACH). In comparison with the methods outlined previously, ACH requires less sophistication to derive and can be implemented by an organization with limited resources. The ACH methodology is a formal approach for the simultaneous comparison and assessment of competing reasonable hypotheses that allows for drawing of conclusions about their relative likelihood based on collected evidence.[35] The methodology was developed in the 1970s by former Central Intelligence Agency methodology expert Richard Heuer and consists of several steps focused on

---

[32] Ellen Nakashima, "Russian Spies Hacked the Olympics and Tried to Make It Look Like North Korea Did It, U.S. Officials Say," *Washington Post*, February 24, 2018; and Andy Greenberg, "A Brief History of Russian Hackers' Evolving False Flags," *Wired*, October 21, 2018.

[33] Interview with a cybersecurity expert, conducted by Bilyana Lilly, Las Vegas, Nev., August 2019 (name withheld on request).

[34] Leekyung Ko, "North Korea as a Geopolitical and Cyber Actor: A Timeline of Events," blog post, New America, June 6, 2018; Joseph Young, "North Korea Hacked Crypto Exchanges and Ran ICOs to Fund Regime: Report," *CCN*, November 13, 2018; Jim Finkle, "SWIFT Warns Banks on Cyber Heists as Hack Sophistication Grows," Reuters, November 28, 2017; Center for Strategic and International Studies, "Significant Cyber Incidents," webpage, undated; Sean Lyngaas, "UN Report Links North Korean Hackers to Theft of $571 Million from Cryptocurrency Exchanges," *CyberScoop*, March 12, 2019; Chris Bing, "Opsec Fail Allows Researchers to Track Bangladesh Bank Hack to North Korea," *CyberScoop*, April 3, 2017; and United Nations Security Council, "Note by the President of the Security Council," New York, S/2019/171, March 5, 2019.

[35] Simon Pope and Audun Jøsang, "Analysis of Competing Hypotheses Using Subjective Logic," *10th International Command and Control Research and Technology Symposium: The Future of C2, Decision Making and Cognitive Analysis*, January, 2005, pp. 3–4.

identifying fundamental assumptions and listing main hypotheses, incorporation of all relevant data in support of and against each hypothesis, formulating tentative conclusions about the likelihood of each hypothesis, comparing the hypotheses, and identifying indicators for future observation.[36] ACH is useful in predictive analysis because it can facilitate answering both why and when a likely adversary may target the network of a particular organization or type of organization based on observed nontechnical factors, such as geopolitical events or other potential triggers in the environment that may affect the behavior of the adversary. Once information on the adversary's potential motivations is collected, one can filter and analyze relevant data points surrounding hypothesized geopolitical trigger events and derive conclusions about the likely future attack timelines of an adversary, guided by this nontechnical data collected in Step 2, once the strategic motive of a certain threat actor is understood. This likely requires development of an ACH matrix and is intended to allow the defender or CTI analyst to gather as much of an understanding of the adversary as possible. This matrix can include data collected from one's own networks, especially dates on which the respective organization has received spearphishing emails or attacks from an adversary in the past along with the lure themes. Defenders are also advised to conduct open-source research for major geopolitical events related to the nation-state adversary and narrow down several event types hypothesized to be correlated or causal to taskings of their adversary and subsequent phishing campaigns. Recurring event types—or a combination thereof—may comprise initial competing hypotheses that potentially signify nontechnical indicators that could be used to predict a certain cyber incident.

To ensure a holistic examination of all of the relevant evidence that can be provided in support of or against a hypothesis in the ACH methodology from the environment of an adversary, defenders can apply the political, military, economic, social, information, and infrastructure (PMESII) framework. PMESII is a convenient analytical starting point to comprehensively capture operational nontechnical variables that can inform the formulation of hypotheses that are included in ACH. Therefore, PMESII can be applied as a complement to ACH, and as a categorizing grid to ensure a structured approach to incorporating the entire possible set of nontechnical indicators by examining the six major domains from which indicators can originate.[37] According to the abbreviation, *political* includes individuals, events, and institutions related to the formal and informal distribution of governance. The *military* components of PMESII encompass military and paramilitary forces and capabilities, as well as relevant military events. The *economic* aspect focuses on individuals, behavior, institutions, and events focused on the production, distribution, and consumption of resources and services.

---

[36] Pope and Jøsang, 2005, p. 3; and Richard J. Heuer, Jr., *Psychology of Intelligence Analysis*, Washington, D.C.: Central Intelligence Agency, Center for the Study of Intelligence, 1999, pp. 95–96.

[37] Brian M. Ducote, *Challenging the Application of PMESII-PT in a Complex Environment*, Fort Leavenworth, Kan.: United States Army Command and General Staff College, May 2010, p. 3.

Social, or sociocultural, incorporates religious, cultural, and ethnic dynamics within the operational environment. *Information* covers the structures, entities, capabilities, individuals, and processes involved in the collection, dissemination, and processing of information. The *infrastructure* category includes the basic installations, facilities, and processes required for the uninterrupted functioning of a society.[38]

These six categories can facilitate the collation of relevant information, the categorization of data collection domains, and associated methods used to analyze the data into logical compartments, enabling identification, monitoring, and analysis of indicators that are relevant to the behavior of certain cyber threat actors that can contribute to the prediction of these actors' behavior. In particular, each of the six categories can serve as a collection for nontechnical sources originating from the operational environment and the cyber actors' characteristics relevant for each PMESII area. For example, in the political category, defenders can focus on an election as an event to be monitored that can signal a potential increase in malware activity against the organizations in the country where the election is being held. As NATO Cooperative Cyber Defence Centre of Excellence Ambassador Kenneth Geers illustrates through analysis of a global data set of malware traffic, malware spikes have been observed before election cycles in Turkey and the United States, among others. These particular attacks commence with computer and human reconnaissance through applications, proceed with targeted malware dissemination via worm, and conclude with a variety of operations from the passive collection of intelligence to possible manipulation of votes via remote access Trojan.[39] Thus, monitoring national election cycles by collecting information about these events through media and government websites can find indicators that can alert certain organizations in the country holding the election that they can expect an increase in malware activity. Specifically, given the past record of Russian APT phishing campaigns and intrusions into U.S. organizations and structures involved in the conduct of elections, cyber defenders working with U.S. election administrators and election campaign officials may choose to pay specific attention to the TTPs of these APTs during U.S. election cycles. The final chapter of this report illustrates how ACH can be applied to an information environment in combination with PMESII to increase the analytical accuracy of predicting the adversarial behavior of a state-sponsored APT.

Step 2 of SWARM offers an approach to incorporate nontechnical alongside technical indicators into a cyber defense process that can strengthen the ability to predict cyber incidents emanating from state-sponsored cyber threats in particular before the delivery of a weaponized payload to the network of an organization. Such indica-

---

[38] Government of the United Kingdom, Ministry of Defence, *Joint Doctrine Note 4/13: Culture and Human Terrain*, London, JDN 4/13, September 2013; and U.S. Marine Corps, "ASCOPE/PMESII," webpage, undated.

[39] Kenneth Geers and Nadiya Kostyuk, "Hackers Are Using Malware to Find Vulnerabilities in U.S. Swing States. Expect Cyberattacks," *Washington Post*, November 5, 2018.

tors can be derived from an examination of novel data domains and the application of emerging methodologies and techniques that require an organization with a moderately mature cyber defense capability but without the dependency of a large budget. A simplified starting point for this type of organization is the application of the ACH matrix in combination with the PMESII model, which refines methodological rigor by establishing analytical comprehensiveness. In Step 3 of SWARM, defenders use the collected technical CTI to model behavior of their adversaries identified in Step 1 and apply a threat model, such as MITRE's PRE-ATT&CK and ATT&CK, to design a playbook of adversary techniques to emulate against their own network in Step 4 to enhance resilience ahead of the predicted attack by testing and tuning defenses.

CHAPTER SIX

SWARM Step Three: Apply a Threat Model

By this step of SWARM, network defenders should have an idea of who might be targeting their organization, as determined by Step 1. Step 2 should have yielded both nontechnical geopolitical indicators of cyber incidents and technical CTI information about the most likely adversaries, their infrastructure, and their malware. The goal of Step 3 in SWARM is for network defenders to break down or model their adversaries' offensive cyber operations behavior into its finite constituent tactics and techniques along Lockheed Martin's Cyber Kill Chain. This will enable the defender to begin heatmapping adversary techniques and overlaying known existing countermeasures to identify *relevant* visibility gaps against the adversaries' techniques and overall playbooks.

As an output of Step 2, the greater the accuracy in prediction of *when*—and ideally against *which* data—impending cyber incidents or attempts will be perpetrated, the better the prioritization of countermeasure actions and resource allocations within the given time frame can be.

MITRE's PRE-ATT&CK and ATT&CK threat modeling frameworks are the examples used as the threat model for SWARM,[1] but, in this step, defenders could substitute any other threat model if they choose to do so. MITRE's PRE-ATT&CK and ATT&CK threat models are *cyber attack lifecycle* models and were chosen as the preferred threat modeling frameworks because they have caught the most momentum in the industry and are now ubiquitously used and integrated into myriad community-driven open-source tools and resources. But, most importantly, these models are quick and effective to understand and apply. Their success has also been attributed in large part to ATT&CK creating a common language that both offense/red team and defense/blue team can speak. The frameworks help organizations move from *Can we be hacked?*—where the answer is always yes—to *What can we prevent or detect?*

PRE-ATT&CK covers adversary tactics and techniques prior to initial access, which corresponds to the *delivery of a weaponized payload* stage of Lockheed Martin's Cyber Kill Chain (Stage 3). ATT&CK begins with initial access (see Figure 6.1).

---

[1] MITRE | ATT&CK, 2019d; and MITRE | ATT&CK, "Enterprise Matrix," webpage, last modified July 2, 2020g.

Figure 6.1
PRE-ATT&CK and ATT&CK Relative to the Original Cyber Kill Chain

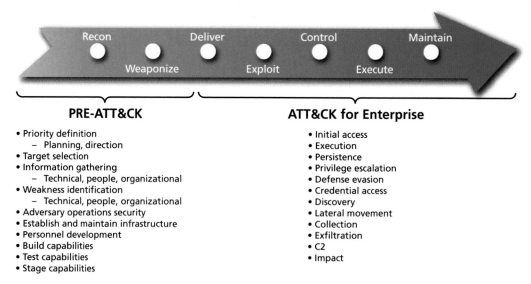

PRE-ATT&CK

- Priority definition
  - Planning, direction
- Target selection
- Information gathering
  - Technical, people, organizational
- Weakness identification
  - Technical, people, organizational
- Adversary operations security
- Establish and maintain infrastructure
- Personnel development
- Build capabilities
- Test capabilities
- Stage capabilities

ATT&CK for Enterprise

- Initial access
- Execution
- Persistence
- Privilege escalation
- Defense evasion
- Credential access
- Discovery
- Lateral movement
- Collection
- Exfiltration
- C2
- Impact

SOURCE: MITRE | ATT&CK, "PRE-ATT&CK Introduction," webpage, undated c.

ATT&CK details the finite number of attacker techniques available, and defenders can match corresponding defensive countermeasures to them in an attempt to break Lockheed Martin's Cyber Kill Chain and stop the incident. See Tables 6.1 and 6.2 for an example of the versions of the PRE-ATT&CK and ATT&CK framework, current as of this writing. Notice that the column heads are sequential tactics from left to right as they go across the Cyber Kill Chain or form a more nuanced evolution of Cyber Kill Chain, beginning with initial access—equivalent to a successful delivery phase—with rows of techniques under each column.

To get started quickly familiarizing oneself with the matrices and the mappings of different APT groups, it is worth experimenting with the web-based ATT&CK Navigator,[2] which can also be hosted locally in one's environment.[3] By selecting the "Prepare" stage in the "Filters" dropdown menu on the website, defenders can access and add components of the PRE-ATT&CK framework.

Some PRE-ATT&CK data, such as the Establish and Maintain Infrastructure tactic (TA0022) and its Buy Domain Name (T1328) technique, may be available from pivoting on IOCs and enumerating adversary infrastructure via passive DNS records in Step 2. Uncovering newly registered domains that are part of adversary C2 infrastructure—especially when they do not yet resolve to an IP address—enables

---

[2]   MITRE Corporation, "MITRE ATT&CK Navigator," webpage, undated.

[3]   "Web App That Provides Basic Navigation and Annotation of ATT&CK Matrices," webpage, GitHub, 2018.

Table 6.1
Snippet of MITRE's PRE-ATT&CK Matrix: PRE-ATT&CK Represents Pre–Initial Access in the Cyber Kill Chain

Priority Definition Planning (13 techniques)	Priority Definition Direction (4 techniques)	Target Selection (5 techniques)	Technical Information Gathering (20 techniques)	People Information Gathering (11 techniques)	Organizational Information Gathering (11 techniques)	Technical Weakness Identification (9 techniques)	People Weakness Identification (3 techniques)	Organizational Weakness Identification (6 techniques)	Adversary OPSEC (20 techniques)	Establish & Maintain Infrastructure (16 techniques)	Persona Development (6 techniques)	Build Capabilities (11 techniques)	Test Capabilities (7 techniques)	Stage Capabilities (6 techniques)
Assess current holdings, needs, and wants	Assign KITs, KIQs, and/or intelligence requirements	Determine approach/attack vector	Acquire OSINT data sets and information	Acquire OSINT data sets and information	Acquire OSINT data sets and information	Analyze application security posture	Analyze organizational skillsets and deficiencies	Analyze business processes	Acquire and/or use 3rd party infrastructure services	Acquire and/or use 3rd party infrastructure services	Build social network persona	Build and configure delivery systems	Review logs and residual traces	Disseminate removable media
Assess KITs/KIQs benefits	Receive KITs/KIQs and determine requirements	Determine highest level tactical element	Conduct active scanning	Aggregate individual's digital footprint	Conduct social engineering	Analyze architecture and configuration posture	Analyze social and business relationships, interests, and affiliations	Analyze organizational skillsets and deficiencies	Acquire and/or use 3rd party software services	Acquire and/or use 3rd party software services	Choose pre-compromised mobile app developer account credentials or signing keys	Build or acquire exploits	Test ability to evade automated mobile application security analysis performed by app stores	Distribute malicious software development tools
Assess leadership areas of interest	Submit KITs, KIQs, and intelligence requirements	Determine operational element	Conduct passive scanning	Conduct social engineering	Determine 3rd party infrastructure services	Analyze data collected	Assess targeting options	Analyze presence of outsourced capabilities	Acquire or compromise 3rd party signing certificates	Acquire or compromise 3rd party signing certificates	Choose pre-compromised persona and affiliated accounts	C2 protocol development	Test callback functionality	Friend/Follow/Connect to targets of interest
Assign KITs/KIQs into categories	Task requirements	Determine secondary level tactical element	Conduct social engineering	Identify business relationships	Determine centralization of IT management	Analyze hardware/software security defensive capabilities		Assess opportunities created by business deals	Anonymity services	Buy domain name	Develop social network persona digital footprint	Compromise 3rd party or closed-source vulnerability/exploit information	Test malware in various execution environments	Hardware or software supply chain implant
Conduct cost/benefit analysis		Determine strategic target	Determine 3rd party infrastructure services	Identify groups/roles	Determine physical locations	Analyze organizational skillsets and deficiencies		Assess security posture of physical locations	Common, high volume protocols and software	Compromise 3rd party infrastructure to support delivery	Friend/Follow/Connect to targets of interest	Create custom payloads	Test malware to evade detection	Port redirector
Create implementation plan			Determine domain and IP address space	Identify job postings and needs/gaps	Dumpster dive	Identify vulnerabilities in third-party software libraries		Assess vulnerability of 3rd party vendors	Compromise 3rd party infrastructure to support delivery	Create backup infrastructure	Obtain Apple iOS enterprise distribution key pair and certificate	Create infected removable media	Test physical access	Upload, install, and configure software/tools
Create strategic plan			Determine external network trust dependencies	Identify people of interest	Identify business processes/tempo	Research relevant vulnerabilities/CVEs			Data Hiding	Domain registration hijacking		Discover new exploits and monitor exploit-provider forums	Test signature detection for file upload/email filters	
Derive intelligence requirements			Determine firmware version	Identify personnel with an authority/privilege	Identify business relationships	Research visibility gap of security vendors			Dynamic DNS	Dynamic DNS		Identify resources required to build capabilities		
Develop KITs/KIQs			Discover target logon/email address format	Identify sensitive personnel information	Identify job postings and needs/gaps	Test signature detection			Host-based hiding techniques	Install and configure hardware, network, and systems		Obtain/re-use payloads		
Generate analyst intelligence requirements			Enumerate client configurations	Identify supply chains	Identify supply chains				Misattributable credentials	Obtain booster/stressor subscription		Post compromise tool development		
Identify analyst level gaps			Enumerate externally facing software applications, languages, and dependencies	Mine social media	Obtain templates/branding materials				Network-based hiding techniques	Procure required equipment and software		Remote access tool development		
Identify gap areas			Identify job postings and needs/gaps						Non-traditional or less attributable payment options	Shadow DNS				
Receive operator KITs/KIQs tasking			Identify security defensive capabilities						Obfuscate infrastructure	SSL certificate acquisition for domain				
			Identify supply chains						Obfuscate operational infrastructure	SSL certificate acquisition for trust breaking				
			Identify technology usage patterns						Obfuscate or encrypt code	Use multiple DNS infrastructures				
			Identify web defensive services						Obfuscation or cryptography					
			Map network topology						OS-vendor provided communication channels					
			Mine technical blogs/forums						Private whois services					
			Obtain domain/IP registration information						Proxy/protocol relays					
			Spearphishing for information						Secure and protect infrastructure					

SOURCE: MITRE | ATT&CK, 2019c.

Table 6.2
Snippet of MITRE's Enterprise ATT&CK Matrix: Begins with Initial Access (Successful Delivery Mechanisms) in the Cyber Kill Chain

Reconnaissance (10 techniques)	Resource Development (6 techniques)	Initial Access (9 techniques)	Execution (10 techniques)	Persistence (18 techniques)	Privilege Escalation (12 techniques)	Defense Evasion (37 techniques)	Credential Access (15 techniques)	Discovery (25 techniques)	Lateral Movement (9 techniques)	Collection (17 techniques)	Command and Control (16 techniques)	Exfiltration (9 techniques)	Impact (13 techniques)
Active Scanning (2)	Acquire Infrastructure (6)	Drive-by Compromise	Command and Scripting Interpreter (8)	Account Manipulation (4)	Abuse Elevation Control Mechanism (4)	Abuse Elevation Control Mechanism (4)	Brute Force (4)	Account Discovery (4)	Exploitation of Remote Services	Archive Collected Data (3)	Application Layer Protocol (4)	Automated Exfiltration (1)	Account Access Removal
Gather Victim Host Information (4)	Compromise Accounts (2)	Exploit Public-Facing Application	Exploitation for Client Execution	BITS Jobs	Access Token Manipulation (5)	Access Token Manipulation (5)	Credentials from Password Stores (3)	Application Window Discovery	Internal Spearphishing	Audio Capture	Communication Through Removable Media	Data Transfer Size Limits	Data Destruction
Gather Victim Identity Information (3)	Compromise Infrastructure (6)	External Remote Services	Inter-Process Communication (2)	Boot or Logon Autostart Execution (12)	Boot or Logon Autostart Execution (12)	BITS Jobs	Exploitation for Credential Access	Browser Bookmark Discovery	Lateral Tool Transfer	Automated Collection	Data Encoding (2)	Exfiltration Over Alternative Protocol (3)	Data Encrypted for Impact
Gather Victim Network Information (6)	Develop Capabilities (4)	Hardware Additions	Native API	Boot or Logon Initialization Scripts (5)	Boot or Logon Initialization Scripts (5)	Deobfuscate/Decode Files or Information	Forced Authentication	Cloud Infrastructure Discovery	Remote Service Session Hijacking (2)	Clipboard Data	Data Obfuscation (3)	Exfiltration Over C2 Channel	Data Manipulation (3)
Gather Victim Org Information (4)	Establish Accounts (2)	Phishing (3)	Scheduled Task/Job (6)	Browser Extensions	Create or Modify System Process (4)	Direct Volume Access	Forge Web Credentials (2)	Cloud Service Dashboard	Remote Services (6)	Data from Cloud Storage Object	Dynamic Resolution (3)	Exfiltration Over Other Network Medium (1)	Defacement (2)
Phishing for Information (3)	Obtain Capabilities (6)	Replication Through Removable Media	Shared Modules	Compromise Client Software Binary	Domain Policy Modification (2)	Domain Policy Modification (2)	Input Capture (4)	Cloud Service Discovery	Replication Through Removable Media	Data from Configuration Repository (2)	Encrypted Channel (2)	Exfiltration Over Physical Medium (1)	Disk Wipe (2)
Search Closed Sources (2)		Supply Chain Compromise (3)	Software Deployment Tools	Create Account (3)	Event Triggered Execution (15)	Execution Guardrails (1)	Man-in-the-Middle (2)	Domain Trust Discovery	Software Deployment Tools	Data from Information Repositories (2)	Fallback Channels	Exfiltration Over Web Service (2)	Endpoint Denial of Service (4)
Search Open Technical Databases (5)		Trusted Relationship	System Services (2)	Create or Modify System Process (4)	Exploitation for Privilege Escalation	Exploitation for Defense Evasion	Modify Authentication Process (4)	File and Directory Discovery	Taint Shared Content	Data from Local System	Ingress Tool Transfer	Scheduled Transfer	Firmware Corruption
Search Open Websites/Domains (2)		Valid Accounts (4)	User Execution (2)	Event Triggered Execution (15)	Hijack Execution Flow (11)	File and Directory Permissions Modification (2)	Network Sniffing	Network Service Scanning	Use Alternate Authentication Material (4)	Data from Network Shared Drive	Multi-Stage Channels	Transfer Data to Cloud Account	Inhibit System Recovery
Search Victim-Owned Websites			Windows Management Instrumentation	External Remote Services	Process Injection (11)	Hide Artifacts (7)	OS Credential Dumping (8)	Network Share Discovery		Data from Removable Media	Non-Application Layer Protocol		Network Denial of Service (2)
				Hijack Execution Flow (11)	Scheduled Task/Job (6)	Hijack Execution Flow (11)	Steal Application Access Token	Network Sniffing		Data Staged (2)	Non-Standard Port		Resource Hijacking
				Implant Container Image	Valid Accounts (4)	Impair Defenses (7)	Steal or Forge Kerberos Tickets (4)	Password Policy Discovery		Email Collection (3)	Protocol Tunneling		Service Stop
				Office Application Startup (6)		Indicator Removal on Host (6)	Steal Web Session Cookie	Peripheral Device Discovery		Input Capture (4)	Proxy (4)		System Shutdown/Reboot
				Pre-OS Boot (5)		Indirect Command Execution	Two-Factor Authentication Interception	Permission Groups Discovery (3)		Man in the Browser	Remote Access Software		
				Scheduled Task/Job (6)		Masquerading (6)	Unsecured Credentials (7)	Process Discovery		Man-in-the-Middle (2)	Traffic Signaling (1)		
				Server Software Component (3)		Modify Authentication Process (4)		Query Registry		Screen Capture	Web Service (3)		
				Traffic Signaling (1)		Modify Cloud Compute Infrastructure (4)		Remote System Discovery		Video Capture			
				Valid Accounts (4)		Modify Registry		Software Discovery (1)					
						Modify System Image (2)		System Information Discovery					
						Network Boundary Bridging (1)		System Network Configuration Discovery					
						Obfuscated Files or Information (5)		System Network Connections Discovery					
						Pre-OS Boot (5)		System Owner/User Discovery					
						Process Injection (11)		System Service Discovery					
						Rogue Domain Controller		System Time Discovery					
						Rootkit		Virtualization/Sandbox Evasion (3)					
						Signed Binary Proxy Execution (11)							
						Signed Script Proxy Execution (1)							
						Subvert Trust Controls (4)							
						Template Injection							
						Traffic Signaling (1)							
						Trusted Developer Utilities Proxy Execution (1)							
						Unused/Unsupported Cloud Regions							
						Use Alternate Authentication Material (4)							
						Valid Accounts (4)							
						Virtualization/Sandbox Evasion (3)							
						Weaken Encryption (2)							
						XSL Script Processing							

SOURCE: MITRE | ATT&CK, 2019c.

early warning of an attack. If the adversary has established a pattern of C2 domain registration and then waiting a certain amount of time before resolution or usage, analysis of the domain registration date relative to the current date may also validate or increase the fidelity of one's predicted window of probable attack. Often, adversaries will register domains intended for C2 and let them sit for 30-plus days to evade a target organization's proxy server settings, which may be set to automatically block newly registered domains because they may be risky.

The IOCs uncovered from this process can be shared with trusted peer organizations and act as early warning for other organizations as well. These indicators are operationalized by incorporating them into an organization's defensive posture for blocking in a proxy server, for example, increasing resilience along with the probability that defenders will break the Cyber Kill Chain prior to the C2 stage—or even earlier—if the attack is broken into multiple delivery stages.

Although adversaries continuously evolve their techniques, it is worth checking whether the adversary has already been modeled with the ATT&CK framework, which will save time. SWARM is effective primarily when applied to state-sponsored threat classes. It is, nevertheless, worth remembering that although these APTs often include the most–highly capable threat actors, these actors do not always employ sophisticated TTPs in all their cyber operations. Some of these state-sponsored intrusion sets, such as Barium (Microsoft), are known to employ certain unsophisticated TTPs, including malicious shortcut files with concealed payloads or basic malware installation vectors in their spearphishing campaigns.[4] Hence, chief information security officers (CISOs) and defenders should bear in mind that, although their organization could be targeted by a nation-state threat actor, they will not necessarily have to defend their networks against zero-day exploits or custom malware, which is difficult to detect.

For the best time efficiency, defenders may implement the following sequence:

1. Check to see whether MITRE already has the respective adversary modeled and mapped to ATT&CK.[5] In this case, Step 3 is completed, unless defenders believe there are additional techniques that have not been mapped.
2. If not, sometimes defenders have reviewed information in Step 2 that may generate some or all of the adversary's TTPs already mapped to the ATT&CK framework, in which case Step 3 is completed.
3. Attempt to find the adversary's ATT&CK mapping in other open-source materials.
4. If the previous steps yield no preexisting mapping, defenders have the opportunity to model their adversary's behavior themselves (and, ideally, share it with the cybersecurity community).

---

[4]   Electronic Transactions Development Agency, 2019, p. 37.

[5]   MITRE | ATT&CK, "Groups," webpage, undated b.

If defenders need to model the adversary themselves by mapping its techniques to PRE-ATT&CK, ATT&CK, or both frameworks, they must ensure that they have familiarized themselves with the techniques in the framework. In this case, defenders are advised to review the technical CTI that they collected in SWARM Step 2. They may be able to clearly map some techniques to their chosen framework immediately, but the more techniques they have mapped under each tactic column, the more of a completely modeled playbook they have for a given adversary. To model their adversaries (if they are not already modeled) or to validate or add to existing ATT&CK mappings, they may wish to review incident data that their organization may have on a given adversary using data from past cyber incidents against the organization.

If the defender is operating within a restricted time frame based on upcoming predicted targeting, defenders may wish to model only the adversary at hand and continue with Step 4. However, it is worth eventually modeling all of the adversaries known to target an organization, prioritized beginning with the ones believed to target the organization's networks within the shortest time period. We recommend modeling them individually and then overlaying the playbooks of each adversary in such a way that defenders can perform frequency analysis and determine which techniques are used most frequently by the adversaries targeting their networks. Focusing on the most frequently used techniques and going down the list will gain defenders the most efficiency in focusing their resources towards improving their countermeasures. Tables 6.3 and 6.4 are some examples of this process.

Step 4—adversary emulation—provides some guidance on how to organize testing networks and endpoints for countermeasure validation or gap discovery via the employment of red team engagements or purple team arrangements.

Table 6.3
A Snippet of APT 35's Playbook Modeled with ATT&CK

Execution	Persistence	Privilege Escalation	Defense Evasion	Credential Access	Discovery	Lateral Movement	Collection
Powershell	Account manipulation	New service	Deobfuscated/decode files or information	Account manipulation	Account discovery	Remote desktop protocol	Data from local system
Scripting	Account creation	Valid accounts	File deletion	Brute force	File and directory discovery	Remote file copy	Email collection
Service execution	External remote services	Web shell	Indicator removal on host	Credential dumping	Network service scanning	Remote services	Input capture
	New service		Masquerading	Input capture	System information discovery	Windows admin shares	Screen capture
	Redundant access		Modify registry	Password filter dynamic link library (DLL)	System owner/user discovery		
	Registry run keys/startup folder		Obfuscated files or information				
	Valid accounts		Redundant access				
	Web shell		Scripting				
			Software packing				
			Timestomp				
			Valid accounts				

NOTE: APT 35 is an Iranian government-sponsored intrusion set. It conducts long-term collection to obtain strategic intelligence. It employs complex social engineering techniques, indicative of the group being potentially well resourced.

Table 6.4
A Snippet of an Example Where Multiple APT Groups Are Modeled with ATT&CK and Overlapping Counts of Techniques Ordered

Technique	Description	APTs and Tools	Total #
T1027	Obfuscated files or information: Adversaries may attempt to make an executable or file difficult to discover or analyze by encrypting, encoding, or otherwise obfuscating its contents on the system or in transit. This is common behavior that can be used across different platforms and the network to evade defenses.	APT 29, APT 28, APT 33, APT 37, Lazarus Group, APT 3, APT 31, APT 35	8
T1064	Scripting: Adversaries may use scripts to aid in operations and perform multiple actions that would otherwise be manual. Scripting is useful for speeding up operational tasks and reducing the time required to gain access to critical resources. Some scripting languages may be used to bypass process monitoring mechanisms by directly interacting with the operating system at an application program interface level instead of calling other programs. Common scripting languages for Windows include VBScript and PowerShell but could also be in the form of command-line batch scripts.	APT 29, APT 28, APT 37, Lazarus Group, APT 3, APT 31, APT 35	7
T1105	Remote file copy: Files may be copied from one system to another to stage adversary tools or other files over the course of an operation. Files may be copied from an external adversary-controlled system through the C2 channel to bring tools into the victim network or through alternate protocols, such as File Transfer Protocol. Files can also be copied over on Mac and Linux with native tools, such as scp, rsync, and sftp.	APT 28, APT 33, APT 37, Lazarus Group, APT 3, APT 31, APT 35	7
T1086	PowerShell: PowerShell is a powerful interactive command line interface (CLI) and scripting environment included in the Windows operating system. Adversaries can use PowerShell to perform a number of actions, including discovery of information and execution of code. Examples include the Start-Process cmdlet which can be used to run an executable, and the Invoke-Command cmdlet, which runs a command locally or on a remote computer.	APT 29, APT 28, APT 33, APT 3, APT 31, APT 35	6
T1060	Registry run keys/startup folder: Adding an entry to the "run keys" in the registry or startup folder will cause the program referenced to be executed when a user logs in. These programs will be executed under the context of the user and will have the account's associated permissions level.	APT 29, APT 33, APT 37, Lazarus Group, APT 3, APT 35	6

Table 6.4—Continued

Technique	Description	APTs and Tools	Total #
T1003	Credential dumping: Credential dumping is the process of obtaining account login and password information, normally in the form of a hash or a clear text password, from the operating system and software. Credentials can then be used to perform lateral movement and access restricted information.	APT 28, APT 33, APT 37, Lazarus Group, APT 3, APT 35	6
T1002	Data compressed: An adversary may compress data (e.g., sensitive documents) that are collected prior to exfiltration to make them portable and minimize the amount of data sent over the network. The compression is done separately from the exfiltration channel and is performed using a custom program or algorithm, or a more common compression library or utility, such as 7zip, RAR, ZIP, or zlib.	APT 28, APT 33, Lazarus Group, APT 3, APT 31, APT 35	6
T1193	Spearphishing attachment: This is a specific variant of spearphishing. Spearphishing attachment is different from other forms of spearphishing in that it employs the use of malware attached to an email. All forms of spearphishing are electronically delivered social engineering targeted at a specific individual, company, or industry. In this scenario, adversaries attach a file to the spearphishing email and usually rely on user execution to gain execution.	APT 29, APT 28, APT 37, Lazarus Group, APT 35	5
T1203	Exploitation for client execution: Vulnerabilities can exist in software due to unsecure coding practices that can lead to unanticipated behavior. Adversaries can take advantage of certain vulnerabilities through targeted exploitation for the purpose of arbitrary code execution. Oftentimes the most valuable exploits to an offensive tool kit are those that can be used to obtain code execution on a remote system because they can be used to gain access to that system. Users will expect to see files related to the applications they commonly use to do work, so they are a useful target for exploit research and development because of their high utility.	APT 29, APT 28, APT 33, APT 37, Lazarus Group	5

Table 6.4—Continued

Technique	Description	APTs and Tools	Total #
T1204	User execution: An adversary may rely on specific actions by a user to gain execution. This may be direct code execution, such as when a user opens a malicious executable delivered via spearphishing attachment with the icon and apparent extension of a document file. It also may lead to other execution techniques, such as when a user clicks on a link delivered via spearphishing link that leads to exploitation of a browser or application vulnerability via exploitation for client execution. While user execution frequently occurs shortly after initial access, it may occur at other phases of an intrusion, such as when an adversary places a file in a shared directory or on a user's desktop hoping that a user will click on it.	APT 29, APT 28, APT 33, APT 37, Lazarus Group	5
T1056	Input capture: Adversaries can use methods of capturing user input for obtaining credentials for valid accounts and information collection that include keylogging and user input field interception. Keylogging: This is the most prevalent type of input capture, with many different ways of intercepting keystrokes, but other methods exist to target information for specific purposes, such as performing a user account control prompt or wrapping the Windows default credential provider.	APT 28, Lazarus Group, APT 3, APT 31, APT 35	5
T1005	Data from local system: Sensitive data can be collected from local system sources, such as the file system or databases of information residing on the system prior to exfiltration. Adversaries will often search the file system on computers they have compromised to find files of interest. They may do this using a CLI, such as cmd, which has functionality to interact with the file system to gather information. Some adversaries may also use automated collection on the local system.	APT 28, APT 37, Lazarus Group, APT 3, APT 35	5
T1043	Commonly used port: Adversaries may communicate over a commonly used port to bypass firewalls or network detection systems and to blend with normal network activity to avoid more-detailed inspection. They may use commonly open ports...	APT 29, APT 33, APT 37, Lazarus Group, APT3	5
T1090	Connection proxy: A connection proxy is used to direct network traffic between systems or act as an intermediary for network communications. Many tools exist that enable traffic redirection through proxies or port redirection, including HTRAN, ZXProxy, and ZXPortMap.	APT 28, APT 3, Lazarus Group, APT 31, APT 35	5

SOURCE: MITRE | ATT&CK, 2019d.

CHAPTER SEVEN

SWARM Step Four: Adversary Emulation

Step 4 of the SWARM model is adversary emulation, sometimes called threat emulation, and its purpose is to enhance resilience. Adversary emulation can be done by a red team or a purple team. Red teams and purple teams ultimately have the same goal: to improve the security posture of the organization by assessing existing defensive countermeasures against the relevant techniques of adversaries who target the organization. This process will validate existing countermeasures and identify relevant gaps in visibility coverage as early as possible ahead of the next incident. Adversary emulation operationalizes the threat modeling done in Step 3, allowing for defensive countermeasure improvement prioritization and resource planning, allocation, or programming, if necessary. SWARM is flexible for usage of either a red or purple arrangement depending on the resources, capabilities, or objectives of an organization. We will first discuss how we consider blue, red and purple teams to be defined to highlight some key differences between them.

Blue team: A blue team is another name for a network defense team. A blue team typically entails the network defenders operating and watching a massive and complex suite of software tools to actively monitor all zones of the network and endpoints—often to a forensic level—in near-real time for anomalous activity, as well as actively hunting for compromises.

Red team: A red team assesses resilience as demonstrated through the blue team's prevention, detection, and response capabilities, as well as by highlighting key network security vulnerabilities that should be prioritized for remediation within the context of the network environment's defense-in-depth architecture. A traditional red team has a penetration testing objective and does not communicate to the blue team during an engagement; anomalous or malicious activity detected by the blue team is validated as real-world or red team activity, and is done via communication through a neutral "white cell" intermediary. Many organizations have annual full-scope red team exercises that are conducted by a third party if the organization does not have a resident red team. A modern red team should be highly technically skilled in the latest offensive techniques, including those in current use by advanced adversaries, and can run an intrusion scenario through the entire Cyber Kill Chain. This is often done by employing general adversary techniques, or by emulating a particular adversary using its unique playbook—a planned-out scenario comprising a sequence of techniques to emulate the behavior of an adversary across the Cyber Kill Chain (see Figure 7.1).

The scope, statement of work, and rules of the red team's engagement are laid out prior to the formal engagement. The final product is often a massive report of findings that are difficult to remediate afterward without a dedicated project manager, and the blue team often benefits only minimally. Another point about the traditional red team versus blue team arrangement is that the success of each team is predicated on the failure of the other, which does not always result in the healthiest environment or the greatest efficiency. However, red teams are still valuable. Additionally, defenders can ask their red team to map adversaries' attacks and techniques to the ATT&CK framework so that the blue team can update its ATT&CK countermeasures heatmap if the final red team report includes this information. The red team's playbook can then be overlaid and compared with that of an organization's adversaries' techniques to help determine which of those techniques the organization can already detect and prevent. This process can also reveal any techniques defenders have not previously identified.

Purple team: Purple teams are appropriate to employ for continuous improvement of resilience. A purple team first and foremost breaks down the wall that exists between the attacker and defender in a traditional red team versus blue team engagement. Purple teams are considered to be blue teams that either possess or develop the ability to perform offensive cyber techniques against their network or an endpoint, or dedicated offensive operators are embedded on the blue team. Another advantage to this approach is that defenders do not need a penetration tester with years of experience to test individual techniques. The most important distinction is that the offensive capability is communicating and co-located with the blue team so that attacks can be conducted and techniques tested with full awareness and coordination in a more agile manner while prevention, detection, and other defenses can be tuned concurrently.

Defenders should work toward creating an overall heatmap of the ATT&CK matrix that represents their cyber defense team's ability to prevent or detect each tech-

Figure 7.1
Example Snippet of Adversary Emulation Using a Playbook of Techniques

Persistence	Privilege Escalation	Defense Evasion	Credential Access	Discovery	Lateral Movement	Execution	Collection	Exfiltration	Command and Control
Accessibility Features	Accessibility Features	Binary Padding	Brute Force	Account Discovery	Application Deployment Software	Command-Line	Automated Collection	Automated Exfiltration	Commonly Used Port
Appinit DLLs	Appinit DLLs	Bypass User Account Control	Credential Dumping	Application Window Discovery	Exploitation of Vulnerability	Execution through API	Clipboard Data	Data Compressed	Communication Through Removable Media
Basic Input/Output System	Bypass User Account Control	Code Signing	Credential Manipulation	File and Directory Discovery	Logon Scripts	Graphical User Interface	Data Staged	Data Encrypted	Custom Command and Control Protocol
Bootkit	DLL Injection	Component Firmware	Credentials in Files	Local Network Configuration Discovery	Pass the Hash	PowerShell	Data from Local System	Data Transfer Size Limits	Custom Cryptographic Protocol
Change Default File Handlers	DLL Search Order Hijacking	DLL Injection	Exploitation of Vulnerability	Local Network Connections Discovery	Pass the Ticket	Process Hollowing	Data from Network Shared Drive	Exfiltration Over Alternative Protocol	Data Obfuscation
Component Firmware	Exploitation of Vulnerability	DLL Search Order Hijacking	Input Capture	Network Service Scanning	Remote Desktop Protocol	Rundll32	Data from Removable Media	Exfiltration Over Command and Control Channel	Fallback Channels
DLL Search Order Hijacking	Legitimate Credentials	DLL Side-Loading	Network Sniffing	Peripheral Device Discovery	Remote File Copy	Scheduled Task	Email Collection	Exfiltration Over Other Network Medium	Multi-Stage Channels
Hypervisor	Local Port Monitor	Disabling Security Tools	Two-Factor Authentication Interception	Permission Groups Discovery	Remote Services	Service Execution	Input Capture	Exfiltration Over Physical Medium	Multiband Communication
Legitimate Credentials	New Service	Exploitation of Vulnerability		Process Discovery	Replication Through Removable Media	Third-party Software	Screen Capture	Scheduled Transfer	Multilayer Encryption
					Windows Management				

SOURCE: Andy Applebaum, "Lessons Learned Applying ATT&CK-Based SOC Assessments," SANS Security Operations Summit, MITRE Corporation, June 24, 2019.

nique, starting with the modeled adversaries targeting their organization. Adversary emulation helps fill out the heatmap with an assessment of the available countermeasure coverage for prioritized offensive techniques, as well as any gaps (see Table 7.1).

If an organization is well resourced, a well-orchestrated combination of all three types of team arrangements would be ideal, but this is rare. Although a red team lends itself well to emulating a well-modeled and complete adversary playbook all at once, a purple team arrangement is better suited for continuous small improvements to prevention and detection against salient and emerging adversary techniques individually; this arrangement, therefore, enhances not only resilience more continuously but also the technical knowledge of the defenders, contrasting with the separation intrinsic to the red team versus blue team arrangement. Additionally, a purple team arrangement is useful (1) if an organization has an incomplete playbook of an adversary who targets it, but wants to know defensive coverage for the adversary's techniques, or (2) to test technique changes and updates that an adversary makes as it evolves. We find this to be most efficient for SWARM from the perspective of continuous improvement because this arrangement allows the blue team to be more agile in more rapidly adapting the defensive security posture of the organization to the changing cyber threat landscape. Additionally, both free and commercial tools to facilitate automated testing of techniques in the ATT&CK framework are being developed and shared.

An annual red team assessment can be used to emulate a particular adversary's entire playbook to assess the defense-in-depth architecture and the blue team as part of the scope of work, as well as any other specific objectives.[1] Whether red, purple, or both, an organization should start by assessing prevention and detection coverage of threat-modeled techniques, identifying and prioritizing gaps, tuning defenses, and then reassessing current coverage.

The following are some free, well-known tools to help an organization get started on adversary emulation that are focused on the purple team approach, or could be used in such a manner:

- *Red Canary's Atomic Red Team*: Despite the name, this is a tool for purple team–style testing of specific ATT&CK techniques and is very quick and easy to get started with. Ultimately, one only has to run a CLI command.[2]
- *MITRE's Caldera*: This is an automated adversary emulation system, built on the MITRE ATT&CK framework. Excellent and well documented, it requires a test environment.[3]

---

[1] For an example comprehensive adversary emulation plan detailed for APT 3, see MITRE | ATT&CK, "Adversary Emulation Plans," webpage, undated a.

[2] "Small and Highly Portable Detection Tests Based on MITRE's ATT&CK," webpage, GitHub, last updated January 8, 2020.

[3] "Scalable Automated Adversary Emulation Platform," webpage, GitHub, last updated January 8, 2020.

Table 7.1
Example Heatmap of Defensive Countermeasures and Gaps

Persistence	Privilege Escalation	Defense Evasion	Credential Access	Discovery	Lateral Movement	Execution	Collection	Exfiltration	Command and Control
	DLL Search Order Hijacking		Brute Force	Account Discovery	Windows Remote Management		Automated Collection	Automated Exfiltration	Commonly Used Port
	Legitimate Credentials		Credential Dumping	Application Window Discovery	Third-party Software		Clipboard Data	Data Compressed	Communication Through Removable Media
Accessibility Features		Binary Padding		File and Directory Discovery	Application Deployment Software	Command-Line	Data Staged	Data Encrypted	Custom Command and Control Protocol
AppInit DLLs		Code Signing	Credential Manipulation			Execution through API	Data from Local System	Data Transfer Size Limits	
Local Port Monitor		Component Firmware		Local Network Configuration Discovery	Exploitation of Vulnerability	Graphical User Interface	Data from Network Shared Drive	Exfiltration Over Alternative Protocol	Custom Cryptographic Protocol
New Service		DLL Side-Loading	Credentials in Files		Logon Scripts	InstallUtil			
Path Interception		Disabling Security Tools	Input Capture	Local Network Connections Discovery		PowerShell	Data from Removable Media		Data Obfuscation
Scheduled Task		File Deletion			Pass the Hash	Process Hollowing		Exfiltration Over Command and Control Channel	Fallback Channels
File System Permissions Weakness		File System Logical Offsets	Network Sniffing		Pass the Ticket	Regsvcs/Regasm	Email Collection		
Service Registry Permissions Weakness		Two-Factor Authentication	Network Service Scanning	Remote Desktop Protocol	Regsvr32	Input Capture	Exfiltration Over Other Network Medium	Multi-Stage Channels
Web Shell		Indicator Blocking	Interception		Remote File Copy	Rundll32	Screen Capture		
		Exploitation of Vulnerability		Peripheral Device Discovery	Remote Services	Scheduled Task	Audio Capture	Exfiltration Over Physical Medium	Multiband Communication
Basic Input/Output System	Bypass User Account Control			Permission Groups Discovery	Remote Services	Scripting	Video Capture		Multilayer Encryption
Bootkit	DLL Injection			Process Discovery	Replication Through Removable Media	Service Execution		Scheduled Transfer	Peer Connections
Change Default File Association	Component Object Model Hijacking	Indicator Removal from Tools		Query Registry	Shared Webroot	Windows Management Instrumentation			Remote File Copy
Component Firmware				Remote System Discovery	Taint Shared Content	MSBuild			Standard Application Layer Protocol
Hypervisor		Indicator Removal on Host			Windows Admin Shares				
Logon Scripts		InstallUtil		Security Software Discovery		Execution through Module Load			Standard Cryptographic Protocol
Modify Existing Service		Masquerading		System Information Discovery					
Redundant Access		Modify Registry							Standard Non-Application Layer Protocol
Registry Run Keys / Start Folder		NTFS Extended Attributes		System Owner/User Discovery					
Security Support Provider		Obfuscated Files or Information							Uncommonly Used Port
Shortcut Modification		Process Hollowing		System Service Discovery					Web Service
Windows Management Instrumentation Event Subscription		Redundant Access		System Time Discovery					Data Encoding
		Regsvcs/Regasm							
Winlogon Helper DLL		Regsvr32							
		Rootkit							
		Rundll32							
		Scripting							
		Software Packing							
		Timestomp							
		MSBuild							
Netsh Helper DLL		Network Share Removal							
Authentication Package		Install Root Certificate							
External Remote Services									

SOURCE: Applebaum, 2019.
NOTE: Prevention is represented by green, detection or deception is represented by yellow, and gaps are red.

- *Uber's Metta*: This is an adversary simulation primarily to test host-based instrumentation.[4]
- *Sliver*: This is an advanced implant simulation framework.[5]
- *Covenant*: This is a .NET C2 framework and collaborative red teaming platform.[6]
- *APT Simulator*: Mapped to ATT&CK, this tool drops artifacts on a system and generates traffic to test detection by simulating a compromised endpoint and "takes less than a minute of your time."[7]
- *FlightSim*: This tool is a network-focused suspicious activity emulation.[8]
- *Awesome Threat Detection*: This is "[a] curated list of awesome threat detection and hunting resources."[9]

Part of the output of adversary emulation is implementing new countermeasures to improve resiliency. This includes planning for collection of new analytics where defenders have visibility gaps to drive threat detection and improve defenses.

---

[4] "An Information Security Preparedness Tool to Do Adversarial Simulation," webpage, GitHub, last updated June 19, 2018.

[5] "Adversary Simulation Framework," webpage, GitHub, last updated January 5, 2020.

[6] "Covenant Is a Collaborative .NET C2 Framework for Red Teamers," webpage, GitHub, last updated November 6, 2019.

[7] "A Toolset to Make a System Look as If It Was the Victim of an APT Attack," webpage, GitHub, last updated June 13, 2018.

[8] "A Utility to Generate Malicious Network Traffic and Evaluate Controls," webpage, GitHub, last updated September 18, 2019.

[9] "A Curated List of Awesome Threat Detection and Hunting Resources," webpage, GitHub, May 13, 2019.

CHAPTER EIGHT

Case Study: Applying SWARM to Predict Phishing Campaigns from the North Korea–Nexus Kimsuky Threat Actor

In the past ten years, approximately a dozen intrusion sets from six countries have targeted RAND. This chapter provides a case study on the application of SWARM, based on a series of unsuccessful North Korean cyber intrusions that targeted RAND's information environment in 2018 and 2019. This case study uses data provided by the RAND Cyber Defense Center and applies SWARM to explore whether there are meaningful correlations between certain geopolitical events and subsequent phishing campaigns targeting RAND that have been attributed to the North Korean intrusion set known alternatively as "Kimsuky" (Kaspersky), "Thallium" (Microsoft), or "Velvet Chollima" (CrowdStrike). This case study follows the SWARM steps as another test of the model.[1] Although the operational methodology presented in SWARM Step 1 and the first half of Step 2 is a fairly high-level overview and standard blend of cyber defense operations and CTI in practice, the intent is to highlight certain aspects from the early warning and resilience frameworks discussed in this report. We hope that the SWARM methodology will be tried and adopted by other organizations targeted by the state-sponsored cyber espionage threat class. At the very least, SWARM provides steps that enhance resilience and improve an organization's defensive cybersecurity posture and situational awareness. Additionally, it can potentially help reframe the impact of existing cyber defense teams' actions within the contexts of early warning and resilience.

Step One Applied—Identify Relevant Cyber Adversaries

Organization Type

As an organization type, RAND is unique in that it is a nonprofit multidisciplinary research institution with defense contracts and nondefense contracts, as well as a gradu-

---

[1] The first case study using an earlier iteration of SWARM looked at Russia's APT 29 intrusion set; see Lilly et al., 2019.

ate school. RAND therefore could be defined under several organization types—think tanks and NGOs, educational institutions, and defense industrial base—which means that it has a variety of state-sponsored adversaries.

Considering the types of intrusion sets that target RAND and these adversaries' other targets, cyber intrusion attempts against RAND seem to be initiated primarily because of the perception of RAND as a think tank. RAND is usually targeted simultaneously in the same campaigns as other organizations perceived as think tanks (e.g., the Council on Foreign Relations, the Brookings Institution), supporting its classification in the SWARM framework as a think tank organization type and highlighting the importance of rapid information-sharing with peers and the integration of CTI processes into cyber defense operations. As a strategic policy research institution that also does national security research, RAND is often targeted by the same intrusion sets that target diplomatic entities, such as ministries of foreign affairs of different countries and the UN, and such government entities as DoD, although usually in separate campaigns. Whether an intrusion set targets RAND simultaneously, separately, or in sequence with other organizations depends on the intrusion set and on the other organization types that the intrusion set targets.

Cyber Threat Class

State-sponsored espionage and cybercrime threat classes both target RAND. Cybercrime activity attempts (all unsuccessful) observed consist of

- social engineering: wire fraud attempts
- social engineering: gift card attempts
- malware in phishes: mostly caught by RAND's first and second lines of email security defenses
- malware from the internet: no match for basic layered defenses.

Although cybercrime is continuous and driven by the desire for profit, our observation is that cyber espionage has a different pattern and is not as continuous. Our hypothesis is that, for some intrusion sets in the espionage threat class, there may be geopolitically related patterns yielding some level of predictability. Cyber espionage tradecraft is typically more advanced, better resourced, and more likely to succeed. In this case study, we focus only on cyber threat actors from the state-sponsored threat class and information collection or espionage subclass, which also seems best suited to SWARM.

Cyber Threat Actors

Practicing threat intelligence requires the development of a better understanding of an organization's adversaries, what motivates them, and what TTPs they use. In this case study, the focus is on North Korea's Kimsuky intrusion set (Velvet Chollima

[CrowdStrike], Thallium [Microsoft]) which seems to have begun targeting RAND in 2018. This attribution to Kimsuky was determined to have high confidence in the assessment through alerts from security devices correlated with malware analysis, pDNS analysis, OSINT, and closed-source unclassified reports. The attribution of certain phishing campaigns to Kimsuky was made from unclassified resources and, in the case of most campaigns, there was compelling evidence to establish attribution.

Kimsuky is most likely affiliated with Bureau 121 under the Reconnaissance General Bureau (RGB) of North Korea's State Affairs Commission. This organizational-level attribution of offensive cyber operations to RGB's Bureau 121 has been public since at least 2014.[2] Bureau 121 consists of Lab 110, two numbered units, and two liaison offices (Figure 8.1). Information from Kong, Lim, and Kim (2019) shows that Lab 110 comprises three offices, each of which seem to play a role in assembling pieces of offensive cyber operations. For example, Office 35 concentrates on developing malware and researching and analyzing vulnerabilities, exploits, and hacking tools, while Office 98's mission is to "[c]ollect information on North Korean defectors, organizations that support them, overseas research institutes related to North Korea, and university professors in South Korea."[3]

Though RAND falls directly within the mission scope of Office 98 under this newly reorganized Bureau 121 structure, it is unclear which office or unit conducts the phishing campaigns or intrusions because it is a collaborative effort. Lab 110 seems to be the development effort. It is possible that Bureau 121's Unit 91 is the unit that actually conducts the cyber espionage phishing campaigns and intrusions, but, for this report, we will refer to the intrusion set as Kimsuky.

Step Two Applied—Focus All-Source Intelligence Collection

The identification of threat actors in Step 1 answers the question of *Who is targeting or most likely to target one's information environment?* This can assist with the prioritization of technical CTI collection that focuses on atomic IOCs and TTPs associated with the identified relevant cyber threat actors. IOCs are tactical-level threat intelligence, while TTPs are operational-level threat intelligence.

Step 2 provides an overview of the basic technical CTI collection and operationalization process used to enhance resilience, as well as strategic-level threat intelligence in the form of nontechnical OSINT analysis to answer the questions: *Why*

---

[2] Ju-min Park and James Pearson, "In North Korea, Hackers Are a Handpicked, Pampered Elite," Reuters, December 4, 2014.

[3] Kong Ji Young, Lim Jong In, and Kim Kyoung Gon, "The All-Purpose Sword: North Korea's Cyber Operations and Strategies," in Tomáš Minárik, Siim Alatalu, Stefano Biondi, Massimiliano Signoretti, Ihsan Tolga, and Gábor Visky, eds., *11th International Conference on Cyber Conflict: Silent Battle, Proceedings 2019*, Tallinn: NATO Cooperative Cyber Defence Centre of Excellence, 2019.

Figure 8.1
North Korea's Military Command Structure, 2018

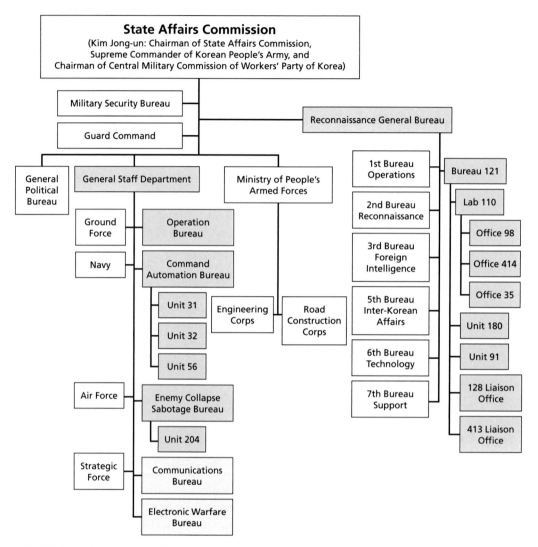

SOURCE: Kong, Lim, and Kim, 2019.

*are they targeting you? What are their goals? What are their objectives and what environ-
mental triggers may exist to indicate an imminent attack or campaign?* More than likely,
review of OSINT and closed-source information—including one's own incident data
and additional analysis performed while operationalizing tactical IOCs—will provide
some answers to the following questions to proceed with predictive analysis:

• Where are the adversary's targets? What is its targeting profile?
• When has it targeted you before and whom has it targeted?

- What are we up against? What are the capabilities of its offensive cyber operators and its malware?

Technical CTI Collection and Operationalization

Once the RAND Cyber Defense Center determined that testing predictability of the Kimsuky intrusion set was a goal because of the high probability of continued spear-phishing campaigns, it began collecting technical CTI reports and IOCs from peer information-sharing forums, OTXs, OSINT, and closed-source unclassified reporting to collect any previously unknown IOCs, TTPs, and adversary playbooks. Malware captured from previous campaigns was re-examined, and reports related to Kimsuky's "BabyShark" malware were collected to derive IOCs and understand its capabilities.

Next, researchers began operationalizing the collected tactical-level threat intelligence to enhance resilience by threat-hunting and infrastructure hardening. Defenders operationalized the IOCs by the following repeatable process that organizations can follow or adapt to their environment.

Document and enrich: IOCs were first imported to a threat intelligence database (the Malware Information Sharing Platform, known as MISP) to determine whether the IOCs autocorrelated to any other events in the database.[4] If an organization is also feeding their intelligence database with various IOCs from their SIEM, any correlations with the new IOCs that are added may also illuminate and attribute prior activity in one's environment. Through a native MISP-PassiveTotal integration, one can choose to enrich IOCs from within MISP with related IOCs or artifacts from PassiveTotal.

Enumerate adversary infrastructure: Once domains and IPs were collected, pDNS was queried to enumerate adversary infrastructure and potentially discover additional related IOCs to supplement existing IOCs, artifacts, and self-generated intelligence from previous encounters with Kimsuky phishing campaigns. Ideally, this step should uncover newly registered infrastructure that has not resolved or been used yet, which can be used to predict which IOCs will be used in the future and premitigate cyber incidents via operationalizing those IOCs and sharing that information with trusted peer organizations. However, although Kimsuky's use of legitimate-but-compromised websites for the first stages of infection, as well as C2, made pDNS less useful for uncovering any newly registered infrastructure in this case, PassiveTotal still revealed that the legitimate-but-compromised domains had malware hashes associated with them via VirusTotal, Hybrid-Analysis.com, and other integrations.

Although no tactical-level predictive indicators emerged from attempting to enumerate adversary threat infrastructure, connections to additional malware hashes and

---

[4] "MISP (Core Software)—Open Source Threat Intelligence and Sharing Platform (Formerly Known as Malware Information Sharing Platform)," webpage, GitHub, undated.

VirusTotal automated dynamic analysis sandbox results were revealed, leading to attribution validation and additional IOCs, such as registry keys, for threat-hunting.

Block or redirect IOCs: IOCs were added for blocking in the various layers of the defense-in-depth architecture to harden the network against future malicious activity. For example, malicious domains were blocked at the proxy, and malware hashes were banned globally via application "allow listing." Note that if a defender's intelligence database is feeding certain IOC types into its security infrastructure, blocking of the IOCs will be completed as soon as the IOCs are scheduled to push. If not, defenders will need to manually add them. In Step 3, searching and threat-hunting, IOCs were added to the SIEM for alerting; even though they should be blocked, defenders should still know if a system tries to execute a payload or reach out to C2 infrastructure.

Email addresses, instead of blocking, were redirected to a defender-controlled mailbox to disrupt the delivery stage of the Cyber Kill Chain while alerting defenders of any subsequent campaigns from the same senders, since Kimsuky often reuses email addresses. The redirection of email enables immediate notification and visibility into subsequent campaigns, which has proven valuable for Kimsuky's typical subsequent campaign—which targeted the same individual for a second time a few days later—and for all campaigns beginning in May 2019.

Searching and threat-hunting: IOCs were added to filters in a SIEM for automated correlation with real-time traffic, retroactive activity searching, and future alerting. Threat-hunting was also performed across the enterprise for such indicators as registry keys, known to be set by the adversary's malware (see Figure 8.2 for one example).[5]

Enterprise patching prioritization: While collecting CTI, analysts should also look out for common vulnerabilities and exposures (CVEs)—known to be exploited by their adversaries—to prioritize patching within their organization. In Kimsuky's case, however, research revealed that, like most modern adversaries, this intrusion set does not usually exploit vulnerabilities, but instead "live off the land." This begins with a spearphishing message enticing the recipient to enable macros on the Microsoft Word document that is phished in or retrieved after clicking on a hyperlink, opening the door and starting a scripted chain of events.

Figure 8.2
Enterprise-Wide Threat-Hunting for One of the Registry Keys Discovered During Malware Analysis

RETURN	Hostname		
WHERE	Host Set *equals* **All hosts**	AND	Registry Key Full Path *contains* **\REGISTRY\MACHINE\SYSTEM\ControlSet001\Services\tunnel\Enum**

SOURCE: RiskIQ, undated. Used with permission.

---

[5] The malware that is referred to here is known as LATEOP (FireEye) and BabyShark (Palo Alto Networks, Unit 42).

Reports that we reviewed also yielded TTPs that are useful for modeling the threat in Step 3, beyond what had been observed in previous campaigns against RAND. This provided an expanded understanding of the adversary's behavioral playbook—which can then be used to emulate Kimsuky's techniques against one's own organization in Step 4—to test and improve prevention, detection, and countermeasures; to discover visibility gaps; and to continue to further enhance resilience against future attacks. In the next, nontechnical part of Step 2, predictive analysis was tested.

Nontechnical OSINT Collection and Analysis

Why are they targeting us? What provokes North Korea's tasking of the Kimsuky offensive cyber operators to conduct this activity? Open-source research revealed that, in 2016–2017, there was a reorganization of the State Affairs Commission's RGB, Bureau 121.[6] This, combined with potential new funding allocation from cyber heists conducted by APT 38/Stardust Chollima, is assessed to be the enabler for the increased operational tempo of this intrusion set. This occurred alongside escalating tensions between the United States and North Korea, which were sparked by a variety of North Korea–centric geopolitical events. Within this mix of event types, it may be possible to achieve some level of probabilistic prediction as to when the Kimsuky phishes will target an organization next, if the relevant environmental triggers are identified.

The predictive analysis goal is to answer the question, *When will they target us next?* The answer could be in the form

> *Abc* geopolitical event(s) occurred, signaling significantly increased probability that we will see a phishing campaign from this intrusion set between the next x and y days.

To determine whether this goal is achievable, an experimental methodology was designed. First, we examine the limitations of the data collection and analysis.

Limitations

The number of data points for attack: The causation between events concerning the United States and North Korea and phishing attempts on RAND staff is limited by the number of data points—seven phishing attempts in 15 months. More data points, in the form of attempted cyber intrusions, are needed to enhance the rigor of the presented assessment of Kimsuky's behavioral pattern, but this cannot be controlled by the researchers.

The number of data points for geopolitical events: More dates could be added to increase the confidence of conclusions. However, because Kimsuky's operational tempo has increased, there are only a limited number of data points that we can use.

---

[6] Kong, Lim, and Kim, 2019.

Over 50 geopolitical events were analyzed in total, and 38 fit the scope of the data point categories.

Date tolerance: Some dates for geopolitical events uncovered during research may have a tolerance of plus or minus one day because of conflicting reporting and relative time zones; the conflicting reporting is because of the isolated nature of the North Korean regime. However, every effort was made to be accurate. Additionally, the time zone difference between the United States (and, in the case of the August 2019 geopolitical events, Europe) and North Korea may contribute to a delay between the period in which North Korea's leader Kim Jong-un receives news and the time when he is likely to factor them into a decision. Such differences may contribute to a one-day lag not reflected in the precise measurement of the time delta between certain events and initiation of cyber incidents.[7]

Misattribution of cyber incidents: In some cases, other organizations misattributed Kimsuky activity to other North Korean intrusion sets, and in other cases failed to attribute some activity to Kimsuky. In most cases, this issue has been corrected, but in one case, the geopolitical data point of Kimsuky phishes aimed at non-RAND entities had to be discarded because of insufficient information to establish attribution.

Unknown missing data: It is not always known when other organizations are targeted, especially South Korean ones. Therefore, we relied on open-source reporting for information about non-RAND targeting.

Known missing data: Geopolitical dates surrounding a July 2018 RAND credential-harvesting campaign were not sought within the scope of this report thus far because this operation was an outlier and also the first instance of targeting, which was different from subsequent campaigns. More geopolitical data points should be gathered surrounding the non-RAND phishes too.

Methodology

To augment what was learned in Step 1 about a particular threat actor targeting an organization and to help shape one's competing hypotheses for determining the correlation or causality of the attacks or phishing campaigns, we recommend using one's own incident data, combined with CTI and OSINT research, to answer *who, where, when*, and *against whom, why* and *what* questions about one's adversary. The following are some example analytical questions to try to answer. Paying close attention to *why* is at the heart of determining what it is that may be behind the timing of targeting, and an ACH matrix is recommended for a fair evaluation of the hypotheses against the data, combined with PMESII, to capture operational nontechnical variables that can inform the formulation or refinement of hypotheses.

---

[7]   North Korea is Coordinated Universal Time (UTC) +9 hours, while RAND's Santa Monica headquarters (Pacific Daylight Time) is UTC –7 (and UTC –8 during Pacific Standard Time). North Korea does not observe daylight saving time changes. From 2015 to 2018, North Korea also observed a clock offset by 30 minutes to symbolically break from imperialism, declaring Pyongyang Time (PYT).

An overview of the methodology is as follows:

1. Review all sources of intelligence, including internal incident data and nontechnical OSINT, to try to answer the following questions:
   a. Who is the attacker?
   b. Where are the attacker's targets? What is their targeting profile?
   c. When have they targeted us before, and who else have they targeted?
      ○ Plot the dates on which RAND received spearphishing emails from Kimsuky, along with the lure themes. If data exist for targeted non-RAND entities, do the same. Make note of the area of expertise of the targeted individuals to see whether there are any discernible patterns.
   d. What are we up against? What are the capabilities of their offensive cyber operators and their malware?
   e. Why are they targeting us? What are their goals?
   f. When will they target us next?
      ○ Conduct open-source research for major North Korea–related geopolitical events and narrow down several event types (data points) hypothesized to be causal to Kimsuky's taskings and subsequent phishing campaigns, resulting in a list of hypotheses.
2. Plot the events along the entire timeline of the phishes in a spreadsheet.
3. Evaluate results to determine whether there are any meaningful patterns or signals that could be reliable indicators that a phishing campaign in which RAND is targeted may follow. Create a summary timeline plot of those most-probably causal events preceding the RAND phishes, including the time delta between the hypothesized causal events and the phishing campaigns to check for consistent patterns.

Findings

1. Who is the attacker?
   a. See Figure 8.1 for an overview of the North Korean RGB command structure and general attribution assessment of Kimsuky.
2. Where are the attacker's targets? What is its targeting profile?
   a. As of mid-2018, likely as a result of a reorganization of Bureau 121 in 2016–2017 and the allocation of additional funding, Kimsuky's reach is global and no longer confined to South Korea. Figure 8.3 presents countries known to have been targeted as of October 10, 2019. Open-source research seems to indicate that targeting beyond South Korea and the United States began in 2019.

3. When has the attacker targeted us before, and who have they targeted? See Figure 8.4.
   a. July 19, 2018: Ten researchers in RAND's Social and Economic Well-Being division (this was an attempt at credential harvesting).
   b. November 19, 2018: international defense researcher A
   c. March 26, 2019: international defense researcher A
   d. May 28, 2019: research and analysis executive B
   e. June 20, 2019: research and analysis executive B
   f. September 30, 2019: international defense researcher focusing on Asia-Pacific affairs
   g. October 2, 2019: international defense researcher focusing on Asia-Pacific affairs.
4. What are we up against? What are the capabilities of the attacker's offensive cyber operators and their malware?
   a. OSINT was unable to turn up any indications of successful or highly skilled operations conducted by this intrusion set in the United States. This unique intrusion set is persistent but does not currently seem very advanced at bypassing standard defenses during the delivery phase the Cyber Kill Chain as of September 2019. Kimsuky's behavior is modeled in more detail in SWARM Step 3.
5. Why are they targeting us? What are their goals?

Figure 8.3
Seven Countries Targeted by Kimsuky as of October 2019

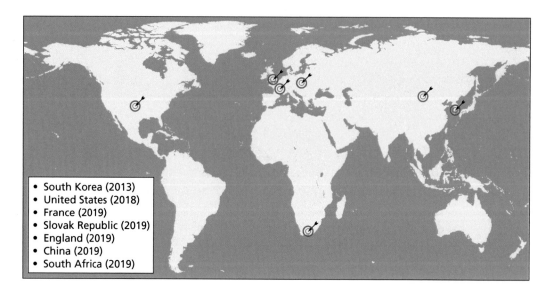

- South Korea (2013)
- United States (2018)
- France (2019)
- Slovak Republic (2019)
- England (2019)
- China (2019)
- South Africa (2019)

Figure 8.4
Timeline of Known RAND Targeting by North Korea's Kimsuky as of October 10, 2019

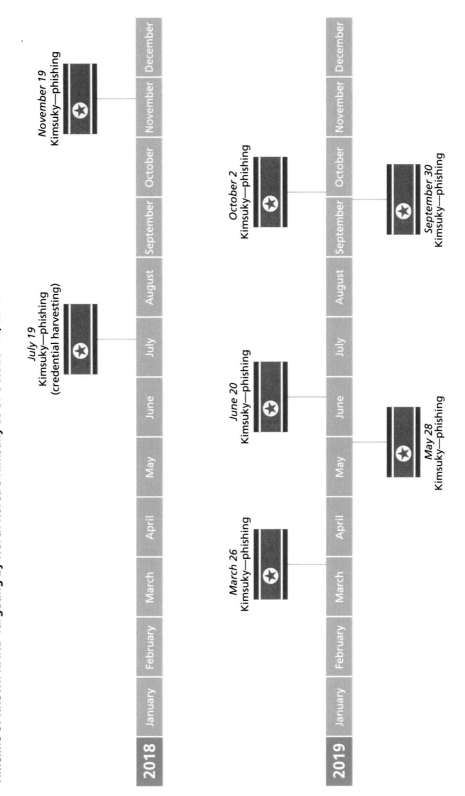

    a. This purpose of this question is to understand the reasoning beyond just the strategic motive of the threat class, such as "espionage or information collection"—the mission of the intrusion set (determined in Step 1 applied)—and additionally formulate hypotheses about which geopolitical event type(s) might be a trigger for activity.

    b. Kimsuky activity extends back to at least 2013, when it started targeting South Korea. We hypothesize that the expansion of cyber targeting beyond South Korea in mid-2018 illustrates a continuing realization of strategic goals and ambition by North Korean leadership made possible, in part, by using money stolen from financial heists and distributed cryptocurrency mining conducted by other hacking units within the North Korean military command structure. Competing hypotheses for increased specificity as to why Kimsuky targets RAND, "as an institute related to North Korea," per the collection requirements mission tasked to Office 98, are

- to measure U.S. response and sentiment following sanctions-related geopolitical events
- to send a political message resulting from dissatisfaction with sanctions-related geopolitical events
- to gain insight into policymaking and policy recommendations coming from think tanks that study North Korean denuclearization, especially in relation to sanctions as a bargaining chip.

The hypotheses at this step will likely remain in play for an organization and be proven, disproven, and refined over time. They are types of questions posed by a CTI or cyber counterintelligence team or function.

6. When will they target us next?

    The following event types, or a combination thereof, are initial competing hypotheses that these observed activities signify nontechnical geopolitical indicators that could be used to predict a phishing campaign.

    These hypotheses have been derived after an application of the PMESII model to the case of Kimsuky. We examined the six areas of PMESII, conducted nontechnical OSINT, and identified the following hypotheses that may indicate an upcoming phishing campaign:

- non-RAND phishes: Known Kimsuky phishes against organizations other than RAND
- U.S. Cyber Command (USCYBERCOM) à la VirusTotal: USCYBERCOM's Cyber National Mission Force uploading North Korean malware to VirusTotal

- US-CERT: US-CERT posting technical malware analysis reports on the HIDDEN COBRA campaign, including reverse-engineering details about North Korean malware, to its website
- sanctions dynamics: This hypothesis covers any new developments related to sanctions—whether new sanctions are imposed, removal of existing ones is denied in negotiations, or violations of existing ones are claimed. The scope of this event type was retroactively expanded from "sanctions imposed" after research began because it became clear that anything sanctions-related seemed to have the same effect, so it made sense to collapse it to one category. The exception was that there were multiple news articles about an ongoing sanctions conversation. This is scoped to "new" sanctions events and discussions.
- indictments: indictments against North Korean hackers
- weapon tests: nuclear, rocket, or submarine missile tests and similar activities
- Trump-Kim meetings: In-person meetings between President Donald Trump and Kim Jong-un.

Table 8.1 is a capture of the raw data of the narrowed geopolitical events plotted, along a timeline, as of October 5, 2019.

Figure 8.5 contains a timeline with what are believed to be the key geopolitical events plotted, in addition to RAND phishes, known non-RAND phishes, and time deltas between the hypothesized correlated or potentially causal event and the phish. Also included is activity in August, for which data were available, involving European sanctions dynamics and a subsequent Kimsuky phish.

Summary of Findings

There seems to be a correlation and possible causation of Kimsuky phishing campaigns that stem from sanctions-related geopolitical events. More data are desired, especially with phishing campaigns; however, we must wait for more phishing campaigns to introduce more data points. RAND set up Google News alerts to detect future North Korean sanctions-related geopolitical events, at which point the general predictive model can be tested for accuracy in predicting or warning of an imminent campaign. Prior to offering the final predictive warning statement, we first review key insights from analysis of the data.

Analysis of the initially hypothesized nontechnical geopolitical indicators in methodology Step 2 yielded some noteworthy insights:

- Non-RAND phishes: Hypothesis rejected. There is no discernible causality or consistent sequence of RAND being either first or last in a cluster of activity during a Kimsuky phishing campaign.
- USCYBERCOM à la VirusTotal: Hypothesis rejected. USCYBERCOM's Cyber National Mission Force began uploading North Korean malware to VirusTotal in September 2019, so there was only one data point. In this case, it was also coordi-

Table 8.1
Screenshot of Raw Data Summary

Date	Phishing Lure Theme	RAND Phish	Non-RAND Phishes	USCYBERCOM --> VT	US-CERT	Sanctions Drama	Indictments	Weapons tests/launches	Trump-Kim Meetings
6/11/18	Unknown		X						
7/19/18	RAND's SEPU	X							
9/6/18							X		
11/12/18									
11/15/18								X	
11/19/18	Nuclear	X							
12/21/18	Foreign Policy; ROK Trip		X						
1/10/19	Cooperation dialogue		X						
2/27-28/2019									X
3/5/19						X			
3/26/19	Nuclear	X							
5/4/19								X	
5/9/19								X	
5/9/19					X				
5/24/19						X			
5/27/19						X			
5/28/19	Nuclear	X	X						
6/11/19						X			
6/18/19	Sanctions renewal		X						
6/19/19	NK missile launch analysis		X						
6/20/19	Nuclear	X							
6/30/19									X
Mid July - Mid August 2019	Unknown		X						
7/25/19								X	
7/31/19								X	
8/2/19						X			
8/2/19								X	
8/16/19								X	
8/17/19								X	
8/20/19	Ballistic Missile Submarine		X						
9/8/19				X					
9/9/19					X				
9/10/19								X	
9/13/19						X			
9/30/19	Trump's UN Speech	X							
10/2/19	Euronews-NBC Interview Request	X							
10/2/19								X	
10/5/19									X

nated with a US-CERT report upload. Later that week, on Friday the 13th, sanctions were imposed on North Korean hackers (a difference of a few days). Moreover, the malware uploaded was not Kimsuky malware. Overall, the hypothesis that USCYBERCOM uploading malware to VirusTotal affected Kimsuky phishing campaigns is rejected.

- US-CERT: Hypothesis rejected. HIDDEN COBRA is a group of North Korean intrusion sets unrelated to the Kimsuky activity, and, so far, US-CERT posting HIDDEN COBRA technical malware analysis reports to its website does not seem to have influenced the timing of Kimsuky phishing campaigns. This is also supported by phishing lure themes relating primarily to sanctions or nuclear issues related to sanctions. US-CERT posted HIDDEN COBRA malware analysis reports on May 9, 2019, and September 9, 2019, 19 days and 17 days prior to Kimsuky phishing campaigns, respectively. At first glance, these events could be taken to be correlated, as this falls perfectly within the same normal distribution of Δt prior to phishing; however, US-CERT HIDDEN COBRA posting dates prior to this definitely had no correlation or causation, and no Kimsuky lure themes have reflected a connection. One thing that was clear, however, is that US-CERT, the Department of Justice, DHS, the Department of the Treasury, and USCYBERCOM prioritize action on North Korean intrusion sets that focus on critical infrastructure and financially motivated attacks rather than espionage

Figure 8.5
Timeline of Sanctions-Related Events Prior to Kimsuky Phishes and Corresponding Time Deltas

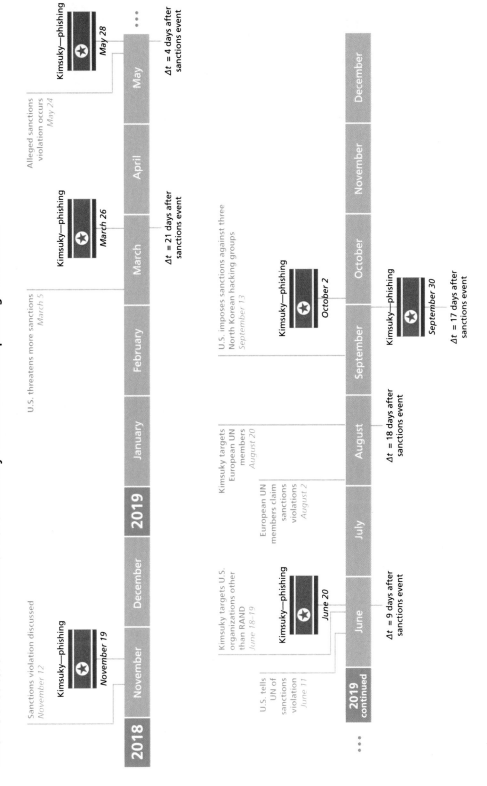

or information collection. Out of the 19 public US-CERT reports that released information regarding North Korean IOCs as of this writing, there is zero overlap with or inclusion of Kimsuky IOCs.

- Sanctions dynamics: Hypothesis accepted. Anything related to sanctions—whether new ones are imposed, removal of existing ones is denied in negotiations, or violations of existing ones are claimed—seems to affect the timing of Kimsuky phishing attacks. This aligns with the regime's desire to continue funding such attacks amid heavy economic sanctions.

It should be noted that the outlier out of the sanctions data points is the May 24, 2019, sanctions event and the May 28, 2019, phish. This is unexplained, but given the later clear seven-day cycle in June (see next paragraph), it is highly unlikely the May 28 phish stemmed from the May 24 event. It was therefore not counted in the median Δt calculations. There could be a missing geopolitical event in early May, or this could have been a planned operation for other unknown reasons and was not necessarily reactionary.

The nuance behind the events leading up to the June 18–20, 2019, phishing campaigns strengthens the sanctions dynamics hypothesis. On June 11, 2019, the United States reported to the UN that North Korean sanctions were being broken: Its oil import cap for the year had already been exceeded. A deadline for UN Security Council members to submit changes to the report in which these claims were made prior to its finalization was set for June 18, which was the day a cluster of phishing campaigns began.[8] This was a seven-day Δt prior to phishing that began after the event, which was likely rushed because the C2 infrastructure would not deliver the second-stage payload. Note that RAND was not targeted until $\Delta t = 9$, so that is what was used in the mean calculations despite phishing campaigns connected to the event occurring at $\Delta t = 7$.

Also of note is the October 2, 2019, phish, two days after the September 30 phish. The adversary's code was found to have an error in it in the September 30 phish: The macro did not properly handle the 302 and 500 HTTP response status codes from the C2. This resulted in Word presenting a macro/Visual Basic for Applications (VBA) error message box, so it did not actually run the "real" part of the macro. The adversary may have realized its code error and tried the phish again on October 2, this time using a Google Drive link for malware downloading.

- Indictments: Hypothesis rejected. Indictments against North Korean hackers only happened once during the time period measured and had no influence on

---

[8]   Pamela Falk, "North Korea Is Sanctions-Busting by Exceeding Oil Import Cap, U.S. Tells UN in New Report," *CBS News*, June 12, 2019.

subsequent phishing campaigns. However, sanctions on hacking groups did have an influence in September 2019.

- Weapon tests: Hypothesis rejected. Weapon tests seem to have less to do with predicting phishing campaigns and more to do with attempting to gain a coercive advantage ahead of meetings with the U.S. President. Phishes seem to be most correlated with sanctions and whether weapon tests are perceived to have been in violation of existing sanctions.
- Trump-Kim meetings: Hypothesis rejected. There was no direct connection between Trump-Kim meetings and Kimsuky phishing campaigns (both before or after the meetings).

Early Warning Conclusions

Most observed time deltas after geopolitical events are either seven to nine days later or 17–21 days later. This likely depends on the perceived urgency of the sanctions-related situation by the North Korean regime. June 2019 was an example of a clearly perceived urgent situation, and there was only a seven-day time delta after the geopolitical event.

The hypothesized reason for high Δt is primarily OSINT collection requirements to develop appropriate targets and lures.[9] Adding to the challenge is that this component is done in the operators' non-native language. However, it is probable that this Δt will decrease over time.

Lure themes seem to mirror what North Korea is interested in learning about. Kimsuky sends highly targeted phishes to carefully selected individuals at carefully selected organizations. This is an area in which Kimsuky shines. Luckily, it has first-stage host-profiling scripts, reliance on macros, and underdeveloped C2 infrastructure, which are all not advanced, so the Cyber Kill Chain is easily broken by common modern security infrastructure.

Despite the current limitations, the research and data analysis seem to suggest that, at least in 2019, there was some level of consistency between sanctions-related points of contention between North Korea and the rest of the world and phishing campaigns originating from Kimsuky. Most phishes are preceded by an open-source news account of sanctions-related dynamics between four and 21 days prior to a phishing campaign, with the mean being 14 days based on the data collected thus far, but seven to nine days and 17–21 days are the two standout temporal distances postevent.

The conclusions lead to the following sample predictive warning statement for Kimsuky phishing campaigns.

A sanctions-related geopolitical event involving North Korea and the United States has occurred, signaling significantly increased probability that we (at RAND)

---

[9] MITRE | ATT&CK, "Acquire OSINT Data Sets and Information," webpage, last modified October 17, 2018a.

will receive a phishing campaign from this intrusion set within a range of seven and 21 days. Temporal warning parameters for Kimsuky are

- probable range: seven to 18 days +/–three days
- mean Δt: 14 days after the triggering event
- median Δt: seven days after the triggering event
- target: If the previous person targeted has only been targeted once by Kimsuky, there is a high probability that they will be the target. If they have already been targeted twice, the future target is unknown.
- mode Δt: seven days after the triggering event.

On October 14, 2019, the RAND Cyber Defense Center developed the above predictive warning statement and started monitoring open-source news events involving North Korean sanctions, using keywords in Google News alerts. There is a near-constant supply of articles written about North Korea that include the keyword "sanctions," but if there is a pattern of geopolitical triggers ahead of phishing campaigns, the significant events will emerge more clearly over time with more data points. On November 10, 2019, the *Washington Times* published an article titled, "Signs Point to Renewed Talks Between U.S., North Korea."[10] The article states,

> U.S. and South Korean officials held a strategy meeting Saturday on the sidelines of the 2019 Moscow Nonproliferation Conference to discuss ways to breathe new life into the talks that broke down last month, and a top official in Seoul said he believes the Trump administration is "very actively" trying to persuade Pyongyang back to the table.

This was a significant geopolitical event that was sanctions-related, even if it was only inferred. As a test, the RAND Cyber Defense Center invoked the predictive analysis statement arrived at in October, estimating that, because the geopolitical event occurred on a Saturday and the article on a Sunday, there would be a significantly increased probability of a phishing campaign by Kimsuky approximately seven to 21 days after Sunday, November 10.[11] This would place the expected phish arrival between November 17 and December 1, 2019. However, the dates came and passed, and no spearphish was detected by the normal mechanisms at the time.

However, CTI-driven investigative leads turned up a spearphish received by one RAND staff member on December 1, 2019. The spearphish sender posed as a South Korean TV writer requesting a Skype interview on the subject of denuclearization

---

[10] Guy Taylor, "Signs Point to Renewed Talks Between U.S., North Korea," *Washington Times*, November 10, 2019.

[11] There is evidence of Kimsuky malware authors working on Sundays (Alyac, "Kimsuky, Operation Stealth Power," blog post, EST Security, April 3, 2019).

negotiations. This spearphish marked a shift in tactics from all previous spearphishes; the difference in this message was the lack of malware, although it did contain a tracking pixel acting as a hidden read receipt. The characteristic of this new tactic indicates a shift to relationship-building prior to transmitting any malicious hyperlinks or documents as an antiresearch and antianalysis technique to better preserve the intellectual property of Kimsuky's malware and infrastructure.

Within the predicted window of time after the geopolitical event, the spearphish arrived 22 days after the event and 21 days after the article (Δt = 21). This falls within the probable range of the temporal warning parameters calculated previously if the plus or minus one day date tolerance mentioned in the "Limitations" section is taken into consideration, given the time it takes for people and information to travel. Though this provides another data point, details on the timing and targeting of other organizations around the same time frame is currently unknown; therefore, the sense of urgency by North Korea derived from the geopolitical event in question that may have precipitated this campaign is difficult to estimate with accuracy, but Δt = 21 is still within the probable range for an expected campaign. Additionally, the adversary demonstrated a change in tactics this time, reflecting potential additional operational planning time.

Kimsuky's operational tempo increased rapidly in 2019, likely because the Bureau 121 reorganization seems to be settled. Behavioral and capabilities evolution of the Kimsuky threat will likely continue to improve and evolve. Sanctions-related geopolitical events seemed to modulate Kimsuky activity in 2019, but because of many international geopolitical dynamics at play for Kim Jong-un and the State Affairs Commission's decisionmaking, the trigger for cyber espionage campaigns may change at any point in time, requiring hypotheses to be adjusted and parts of this process to be repeated to maintain or regain temporal accuracy for early warning. Lastly, Kimsuky is likely to improve turnaround time to a limited degree and reduce its overall turnaround time for target development and operational planning while expanding its C2 infrastructure.

Future Refinements

Link analysis: Using a link-analysis tool, such as Maltego, to map all known Kimsuky C2 infrastructure and understand which hashes are connected to which infrastructure, combined with some knowledge of Kimsuky targeting outside one's own organization, could provide such insights as whether certain infrastructure is used repeatedly for certain targets.

Statistical analysis: Though the data set is not large, additional mathematical modeling and statistical analysis could be applied to the current data and preset to insert future data points as they happen, which may improve the model.

Malware compilation and lure document last-modified timestamps: Analyzing time deltas between BabyShark malware compilation timestamps and when the malware is used in any campaign could yield new insights. A cursory review of malware compile times versus when a sample is actually used shows that malware is often prepared well ahead of time, but more analysis and data points are needed to gain deeper insight into the overall operational tempo of this intrusion set, which has the potential to improve prediction accuracy if other variables in the scenario remain relatively stable. This compilation could also shed light on the hypothesis that the primary delay in campaigns following a relevant event is because of the time necessary for Office 98 to complete technical information-gathering (so that it can select targets if there is already a standard malware authoring pipeline in Office 35) independent of geopolitical events that are driving operational tempo (see Figure 8.1).

Step Three Applied—Apply a Threat Model

As of late October 2019, Kimsuky's playbook has now been modeled by MITRE, though gaps identified from RAND-observed campaigns and technical CTI from OSINT in SWARM Step 2 resulted in an updated playbook. Note that more techniques were observed beyond the MITRE playbook, and they varied slightly across our campaigns as well. An overview of RAND-observed Kimsuky techniques, with a focus on the LATEOP/BabyShark first stage, is as follows, though it did vary slightly across campaigns, so not all were identical to the following example:

- Prior to October 2019, stage 0 was typically a zipped, password-protected macro-laden Microsoft Word file (ATT&CK T1193).[12]
- Enabling VBA macros starts a VBScript to invoke mshta.exe to remotely execute and locally load the stage 1 .hta file (executable malicious HTML) from a legitimate-but-compromised WordPress-based site (Scripting, ATT&CK 106).[13]
- The loaded .hta file sends a request to another part of the C2 server and receives an encoded response, which is decoded locally (ATT&CK T1102).[14]
- The local response adds registry keys to enable future macros (ATT&CK T1058).[15]
- The script then performs a list of CLI commands that correspond to multiple numbered ATT&CK techniques to profile the victim system and saves the results of the system-profiling queries for encoding and exfiltration via an HTTP POST command (ATT&CK TA0007).[16] See Table 8.2 for modeled techniques.

---

[12] MITRE | ATT&CK, "Spearphishing Attachment," webpage, last modified March 27, 2020d.

[13] MITRE | ATT&CK, "Scripting," webpage, last modified June 24, 2019; MITRE | ATT&CK, "Signed Binary Proxy Execution: Mshta," webpage, last modified June 20, 2020f.

[14] MITRE | ATT&CK, "Web Service," webpage, last modified March 26, 2020c.

[15] MITRE | ATT&CK, "Hijack Execution Flow: Service Registry Permissions Weakness," webpage, last modified June 20, 2020e.

[16] MITRE | ATT&CK, "Discovery," webpage, last modified July 19, 2019b.

Malware analysis and cyber intelligence reporting confirmed that the initial mechanism by which Kimsuky's macro-enabled Word document—acting as a VBA-based downloader—interacts with a victim system is to invoke mshta.exe to reach out to the C2 domain to download a .hta file. Mshta.exe is a trusted Windows utility that can be used by adversaries to proxy the execution of malicious .hta files and JavaScript or VBScript. An example of this technique observed in the October 2, 2019, phish can be seen in Figure 8.6.

During automated dynamic analysis, the registry run keys/startup folder technique (T1060) was observed as part of the persistence tactic.[17] Knowing the details of a given technique and how one's adversary uses it can facilitate detection via threat-hunting in the case of registry artifacts (Table 8.3).

Step Four Applied—Adversary Emulation

Once an adversary's tactics, techniques, and behavior are modeled, allowing these adversary tactics to inform defensive actions by emulating them is the last step in SWARM. Testing defensive coverage—prevention, detection, and response—by emulating specific adversary techniques via a purple team arrangement, or emulating an entire playbook via a Red team engagement, enables defensive tuning to close visibility gaps that are discovered during adversary emulation and increases resilience.

Without threat modeling across an adversary's entire playbook prior to having experienced a cyber incident from that adversary, a disadvantage to disrupting the attack so early in the Cyber Kill Chain is a lack of visibility into the adversary's post-exploitation activities. Luckily, Kimsuky's preexisting threat modeling information allows defenders to begin emulation, mitigations, and tuning across the Cyber Kill Chain and playbook while considering the adversary's known techniques and how it can be detected, denied, disrupted, or degraded. This also makes it more costly and less likely for an adversary to completely change its overall behavior or master new techniques. The combination of these activities by defenders increases resilience. We began with a couple of techniques as close to the beginning of the Cyber Kill Chain as possible, in response to the Kimsuky targeting and following our use of SWARM.

ATT&CK technique T1170—malicious usage of mshta.exe—is the first technique (an execution tactic) that we targeted for mitigation. Legitimate usage and need for mshta.exe and .hta files are not common in most enterprises. To mitigate this adversary technique, a metered monitor-only rule in an "application allow listing" solution was set to index and monitor the monitored file's usage for a determined amount of time prior to setting the application "allow listing" tool to index and monitor existing

---

[17] MITRE | ATT&CK, "Boot or Logon Autostart Execution: Registry Run Keys/Startup Folder," webpage, last modified March 25, 2020a.

Table 8.2
Kimsuky's Playbook Threat Modeled by MITRE and Additional RAND Observations

Initial Access	Execution	Persistence	Privilege Escalation
Spearphishing attachment	Mshta	Browser extensions	New service
Spearphishing link	Powershell	Change default file association	Process injection
Scripting		New service	Service registry permissions weakness
Valid accounts		Registry run keys/startup folder	Valid accounts
		Scheduled task	
		Redundant access	
		Valid accounts	

Defense Evasion	Credential Access	Discovery	Lateral Movement
Disabling security tools	Credentials in files	File and directory discovery	Remote desktop protocol
File deletion	Input capture	System information discovery	
Mshta	Credential dumping	Account discovery	
Process injection	Network sniffing	System owner/user discovery	
Redundant access		Process discovery	
Valid accounts		Query registry	
		Network sniffing	

Collection	C2	Exfiltration
Data from local system	Remote access tools	Data encrypted
Input capture	Web service	Exfiltration over C2 channel
	Data encoding	

Figure 8.6
Kimsuky Usage of ATT&CK T1170, Mshta.exe

```
Execute("CreateObject(""WScript.Shell"").Run ""mshta https://███████████████████.hta"", 0, True")
```

NOTE: See MITRE | ATT&CK, 2020g, for more on mshta.exe.

Table 8.3
Example Registry Artifacts

Action	Key	Value
RegSetValue	HKLM\SOFTWARE\Microsoft\Windows NT\CurrentVersion\ Schedule\TaskCache\Tasks\{84663180-3751-4BD9-AF36- 5301A62120D8}\Path	\Adobe\Microsoft\ Windows\qwertd
RegCreateKey	HKLM\SOFTWARE\Microsoft\Windows NT\CurrentVersion\ Schedule\TaskCache\Tree\Adobe\Microsoft\Windows\ qwertd	Not applicable
RegSetValue	HKLM\SOFTWARE\Microsoft\Windows NT\CurrentVersion\ Schedule\TaskCache\Tree\Adobe\Microsoft\Windows\ qwertd\Id	{Mount volume GUID redacted}
RegSetValue	HKLM\SOFTWARE\Microsoft\Windows NT\CurrentVersion\ Schedule\TaskCache\Tree\Adobe\Microsoft\Windows\ qwertd\Index	3

benign execution of this file type for enterprise-wide usage prevalence for a determined amount of time. Next, the application "allow listing" tool was set to block its execution entirely in an application "allow listing" policy. Alerts were subsequently set up for any attempted-but-blocked executions of this file type regardless of parent process, and, tested. This process was followed to deny this ATT&CK T1170 mshta.exe technique during the execution tactic (TA0002) phase right after delivery of a malicious email, when macros are enabled or a malicious link is clicked (in the event that the delivery/ initial access tactics are successful).[18] The ATT&CK execution tactic, using the mshta. exe technique is the stage 0 or *downloader* phase (equivalent to the Cyber Kill Chain's installation phase) prior to retrieving stage 1 of the malware and establishing persistence. Related to this technique—which is used by the BabyShark malware and covers the same execution tactic—is scripted usage of wscript.exe (ATT&CK T1064) to run the malicious payload from a scheduled task once persistence is established.[19] Legitimate Windows script interpreters, such as wscript or cscript, cannot easily be denied from running without negatively affecting an enterprise, so monitoring was verified via EDR alerts and CLI monitoring logs flowing into the SIEM.

The ATT&CK technique T1082—System Information Discovery technique—is used by most adversaries to discover the system and environment in which their malware has landed.[20] For example, Kimsuky has been known to use "ver" on the CLI, as well as other techniques under the discovery tactic (TA0007).[21] Windows CLI moni-

---

[18]   MITRE | ATT&CK, "Execution," webpage, last modified July 19, 2019c.

[19]   MITRE | ATT&CK, "ATT&CK: Scripting," May 31, 2017. Last modified June 24, 2019.

[20]   MITRE | ATT&CK, "System Information Discovery," webpage, last modified March 26, 2020b.

[21]   MITRE | ATT&CK, 2019b.

toring via Windows Event ID 4688, combined with the Group Policy Object setting that expands the Audit Process Creation policy, was verified in the SIEM to ensure that suspicious commands are being logged and monitored, as was successful parsing and ingestion of PowerShell commands into the SIEM for monitoring. Additional Kimsuky tactics and techniques were emulated, and mitigations are being worked toward to ensure coverage and overall resilience. Prioritization of modeled adversary techniques for which to target coverage were sequenced, similar to Table 6.4, for the highest efficiency across the multiple intrusion sets that target RAND.

CHAPTER NINE

Conclusion

In a speech at BlackHat in 2014, the CISO of In-Q-Tel, Dan Geer, stated: "There are . . . parallels between cybersecurity and the intelligence functions, insofar as predicting the future has a strong role to play in preparing your defenses for probable attacks."[1] Geer's comment captures the fundamental assumption of the model presented in this analysis. This report draws on USIC tradecraft and offers a scalable and unique model that cyber defense teams can incorporate into their processes to enhance both the early warning *and* resilience of their defense architecture. SWARM combines technical and strategic indicators to facilitate the prediction, early detection, and warning of cyber incidents targeting its information environment, while also strengthening the resilience of their systems through improving rapid detection and response.

Drawing on warning intelligence frameworks mainly developed and refined by the USIC, the research proposes a four-step scalable threat-centric model that allows defenders to prioritize their resources and focus on protecting their networks against the threat classes and adversaries that are most likely to target them based on their organization type. The model proposes an all-source intelligence collection process that includes both CTI and strategic nontechnical data. Combined with the use of threat modeling frameworks, specifically PRE-ATT&CK and ATT&CK, the model recommends building a comprehensive profile of the cyber adversaries that an organization is likely to face. In the final step, the model advocates periodic red and purple team exercises during which defenders test their systems, assess countermeasure coverage, identify gaps, and update their defenses based on the results of the performed exercises.

This report also demonstrates how the model can be applied in practice by using real-world examples of intrusion attempts into RAND's networks by North Korean cyber threat actor Kimsuky. Using an analysis of both the technical profile and the nontechnical characteristics of Kimsuky allows for the ability to assess with a certain level of probability when Kimsuky is likely to target RAND again. The case study, although tailored to RAND's information environment, illustrates how the steps of the

---

[1] Black Hat, "Cybersecurity as Realpolitik by Dan Geer Presented at Black Hat USA 2014," video, YouTube, August 7, 2014.

model developed in this research may be practically applied by cyber defense teams of different organizations.

SWARM presents a methodological process-based approach for a robust defense architecture that will facilitate prediction and early warning of cyber incidents relative to one's information environment while enhancing resilience, which is especially critical for organizations in the current rapidly evolving cyber threat landscape. As cyber defense teams implement and adapt SWARM for their organizations, we believe that improvements and refinement will result from this operational experience.

References

Abdlhamed, Mohamed, Kashif Kifayat, Qi Shi, and William Hurst, "A System for Intrusion Prediction in Cloud Computing," *Proceedings of the International Conference on Internet of Things and Cloud Computing*, New York, 2016.

Abeshu, Abebe, and Naveen Chilamkurti, "Deep Learning: The Frontier for Distributed Attack Detection in Fog-to-Things Computing," *IEEE Communications Magazine*, Vol. 56, No. 2, February 2018, pp. 169–175.

Accenture, *The Nature of Effective Defense: Shifting from Cybersecurity to Cyber Resilience*, Dublin, 2018. As of January 4, 2020:
https://www.accenture.com/_acnmedia/accenture/conversion-assets/dotcom/documents/local/en/accenture-shifting-from-cybersecurity-to-cyber-resilience-pov.pdf

Aditham, Santosh, and Nagarajan Ranganathan, "A Novel Framework for Mitigating Insider Attacks in Big Data Systems," *Proceedings of the 2015 IEEE International Conference on Big Data (Big Data)*, Santa Clara, Calif., 2015, pp. 1876–1885.

"Adversary Simulation Framework," webpage, GitHub, last updated January 5, 2020. As of January 8, 2020:
https://github.com/BishopFox/sliver

Alyac, "Kimsuky, Operation Stealth Power," blog post [originally in Korean], EST Security, April 3, 2019. As of January 9, 2020:
https://blog.alyac.co.kr/2234

Applebaum, Andy, "Lessons Learned Applying ATT&CK-Based SOC Assessments," SANS Security Operations Summit, MITRE Corporation, June 24, 2019.

AT&T Cybersecurity, "AT&T Alien Labs Open Threat Exchange," webpage, undated. As of January 4, 2020:
https://cybersecurity.att.com/open-threat-exchange

Bellani, Deepak, *The Importance of Business Information in Cyber Threat Intelligence (CTI), the Information Required and How to Collect It*, Bethesda, Md.: SANS Institute Information Security Reading Room, 2017.

Bilge, Leyla, Yufei Han, and Matteo Dell'Amico, "RiskTeller: Predicting the Risk of Cyber Incidents," *Proceedings of the 2017 ACM SIGSAC Conference on Computer and Communications Security*, Dallas, Tex., October 30–November 3, 2017, pp. 1299–1311.

Bing, Chris, "Opsec Fail Allows Researchers to Track Bangladesh Bank Hack to North Korea," *CyberScoop*, April 3, 2017. As of January 4, 2020:
https://www.cyberscoop.com/bangladesh-bank-hack-north-korea-kaspersky/

Black Hat, "Cybersecurity as Realpolitik by Dan Geer Presented at Black Hat USA 2014," video, YouTube, August 7, 2014. As of May 26, 2020: https://www.youtube.com/watch?v=nT-TGvYOBpI&feature=youtu.be

Blackshaw, Amy, "Behavior Analytics: The Key to Rapid Detection and Response?" webpage, RSA, January 22, 2016. As of January 4, 2020: https://www.rsa.com/en-us/blog/2016-01/behavior-analytics-the-key-to-detection-response

Bodeau, Deborah J., Jennifer Fabius-Greene, and Richard D. Graubart, *How Do You Assess Your Organization's Cyber Threat Level?* Bedford, Mass.: MITRE Corporation, August 2010.

Bodeau, Deborah J., and Richard Graubart, *Cyber Resiliency Engineering Framework,* Bedford, Mass.: MITRE Corporation, September 2011.

Busselen, Michael, "CrowdStrike CTO Explains 'Breakout Time'—A Critical Metric in Stopping Breaches [VIDEO]," CrowdStrike, June 6, 2018. As of January 5, 2020: https://www.crowdstrike.com/blog/crowdstrike-cto-explains-breakout-time-a-critical-metric-in-stopping-breaches/

Caralli, Richard A., Julia H. Allen, and David W. White, *CERT Resilience Management Model: A Maturity Model for Managing Operational Resilience*, Boston, Mass.: Addison-Wesley Professional, 2010.

Center for Strategic and International Studies, "Significant Cyber Incidents," webpage, undated. As of January 4, 2020: https://www.csis.org/programs/technology-policy-program/significant-cyber-incidents

Collier, Zachary A., Igor Linkov, Daniel DiMase, Steve Walters, Mark Tehranipoor, and James H. Lambert, "Cybersecurity Standards: Managing Risk and Creating Resilience," *Computer*, Vol. 47, No. 9, September 2014, pp. 70–76.

"Covenant Is a Collaborative .NET C2 Framework for Red Teamers," GitHub, webpage, last updated November 6, 2019. As of January 8, 2020: https://github.com/cobbr/Covenant

"A Curated List of Awesome Threat Detection and Hunting Resources," GitHub, webpage, May 13, 2019. As of January 8, 2020: https://github.com/0x4D31/awesome-threat-detection

Cyber Threat Alliance, "Playbooks," webpage, undated. As of January 4, 2020: https://www.cyberthreatalliance.org/playbooks/

Dalton, Adam, Bonnie Dorr, Leon Liand, and Kristy Hollingshead, "Improving Cyber-Attack Predictions Through Information Foraging," *IEEE International Conference on Big Data (Big Data),* Boston, Mass., December 11–14, 2017.

DeCianno, Jessica, "IOC Security: Indicators of Attack vs. Indicators of Compromise," blog post, CrowdStrike, December 9, 2014. As of January 4, 2020: https://www.crowdstrike.com/blog/indicators-attack-vs-indicators-compromise/

Defense Intelligence Agency, "Warning Fundamentals," unclassified briefing, May 2014.

Department of Defense Directive 3115.16, *The Defense Warning Network*, Washington, D.C.: U.S. Department of Defense, Incorporating Change 2, August 10, 2020.

DHS CISA—*See* U.S. Department of Homeland Security, Cybersecurity and Infrastructure Security Agency.

DIA—*See* Defense Intelligence Agency.

Director of National Intelligence and U.S. Department of Homeland Security, *Cyber Resilience and Response: 2018 Public-Private Analytic Exchange Program*, Washington, D.C., undated. As of January 4, 2020:
https://www.dni.gov/files/PE/Documents/2018_Cyber-Resilience.pdf

Divya, T., and K. Muniasamy, "Real-Time Intrusion Prediction Using Hidden Markov Model with Genetic Algorithm," in L. Padma Suresh, Subhransu Sekhar Dash, and Bijaya Ketan Panigrahi, eds., *Artificial Intelligence and Evolutionary Algorithms in Engineering Systems: Proceedings of ICAEES 2014*, Vol. 1, New Delhi: Springer, India, 2015, pp. 731–736.

DNI and DHS—*See* Director of National Intelligence and U.S. Department of Homeland Security.

Dobrygowski, Daniel, "Cyber Resilience: Everything You (Really) Need to Know," blog post, World Economic Forum, July 8, 2016. As of January 4, 2020:
https://www.weforum.org/agenda/2016/07/cyber-resilience-what-to-know/

DoD—*See* U.S. Department of Defense.

Ducote, Brian M., *Challenging the Application of PMESII-PT in a Complex Environment*, Fort Leavenworth, Kan.: United States Army Command and General Staff College, May 2010. As of January 4, 2020:
https://apps.dtic.mil/dtic/tr/fulltext/u2/a523040.pdf

Duggan, David P., Sherry R. Thomas, Cynthia K. K. Veitch, and Laura Woodard, *Categorizing Threat: Building and Using a Generic Threat Matrix*, Albuquerque, N.M.: Sandia National Laboratory, SAND2007-5791, September 2007.

Eder-Neuhauser, Peter, Tanja Zseby, Joachim Fabini, and Gernot Vormayr, "Cyber Attack Models for Smart Grid Environments," *Sustainable Energy, Grids and Networks*, Vol. 12, December 2017, pp. 10–29.

Electric Power Research Institute, "Grid Resiliency," webpage, undated.

Electronic Transactions Development Agency, *Threat Group Cards: A Threat Actor Encyclopedia*, Bangkok, June 19, 2019. As of January 4, 2020:
https://www.thaicert.or.th/downloads/files/A_Threat_Actor_Encyclopedia.pdf

Falk, Pamela, "North Korea Is Sanctions-Busting by Exceeding Oil Import Cap, U.S. Tells UN in New Report," *CBS News*, June 12, 2019.

Fava, Daniel S., Stephen R. Byers, and Shanchieh Jay Yang, "Projecting Cyberattacks Through Variable-Length Markov Models," *IEEE Transactions on Information Forensics and Security*, Vol. 3, No. 3, September 2008, pp. 359–369.

Finkle, Jim, "SWIFT Warns Banks on Cyber Heists as Hack Sophistication Grows," Reuters, November 28, 2017.

Fuertes, Walter, Francisco Reyes, Paúl Valladares, Freddy Tapia, Theofilos Toulkeridis, and Ernesto Pérez, "An Integral Model to Provide Reactive and Proactive Services in an Academic CSIRT Based on Business Intelligence," *Systems*, Vol. 5, No. 4, 2017, pp. 52–72.

Galinec, Darko, and William Steingartner, "Combining Cybersecurity and Cyber Defense to Achieve Cyber Resilience," *IEEE 14th International Scientific Conference on Informatics*, Poprad, Slovakia, November 14–16, 2017, pp. 87–93.

Geers, Kenneth, and Nadiya Kostyuk, "Hackers Are Using Malware to Find Vulnerabilities in U.S. Swing States. Expect Cyberattacks," *Washington Post*, November 5, 2018.

Government of the United Kingdom, Foreign and Commonwealth Office, National Cyber Security Centre, and Jeremy Hunt, "UK Exposes Russian Cyber Attacks," press release, October 4, 2018. As of January 5, 2020:
https://www.gov.uk/government/news/uk-exposes-russian-cyber-attacks

Government of the United Kingdom, Ministry of Defence, *Joint Doctrine Note 4/13: Culture and Human Terrain*, London, JDN 4/13, September 2013.

Goyal, Palash, Tozammel Hossain, Ashok Deb, Nazgol Tavabi, Nathan Barley, Andrés Abeliuk, Emilio Ferrara, and Kristina Lerman, "Discovering Signals from Web Sources to Predict Cyber Attacks," *arXiv*, preprint, August 2018.

Grabo, Cynthia M., *Warning Intelligence*, McLean, Va.: Association of Former Intelligence Officers, 1987.

Grabo, Cynthia M., *Anticipating Surprise: Analysis for Strategic Warning*, Jan Goldman, ed., Washington, D.C.: Center for Strategic Intelligence Research, Joint Military Intelligence College, December 2002.

Greenberg, Andy, "A Brief History of Russian Hackers' Evolving False Flags," *Wired*, October 21, 2018.

Halawa, Hassan, Matei Ripeanu, Konstantin Beznosov, Baris Coskun, and Meizhu Liu, "An Early Warning System for Suspicious Accounts," *Proceedings of the 10th ACM Workshop on Artificial Intelligence and Security*, Dallas, Tex., November 2017, pp. 51–52.

Hernández, Jarilyn M., Line Pouchard, Jeffrey McDonald, and Stacy Prowell, "Developing a Power Measurement Framework for Cyber Defense," *Proceedings of the Eighth Annual Cyber Security and Information Intelligence Research Workshop*, Oak Ridge, Tenn., January 2013.

Hernandez-Suarez, Aldo, Gabriel Sanchez-Perez, Karina Toscano-Medina, Victor Martinez-Hernandez, Hector Perez-Meana, Jesus Olivares-Mercado, and Victor Sanchez, "Social Sentiment Sensor in Twitter for Predicting Cyber-Attacks Using ℓ1 Regularization," *Sensors*, Vol. 18, No. 5, May 2018, pp. 1380–1397.

Heuer, Richard J., Jr., *Psychology of Intelligence Analysis*, Washington, D.C.: Central Intelligence Agency, Center for the Study of Intelligence, 1999.

Husák, Martin, Jana Komárková, Elias Bou-Harb, and Pavel Čeleda, "Survey of Attack Projection, Prediction, and Forecasting in Cyber Security," *IEEE Communications Surveys and Tutorials*, Vol. 21, No. 1, 2018, pp. 1–21.

Hutchins, Eric M., Michael J. Cloppert, and Rohan M. Amin, "Intelligence-Driven Computer Network Defense Informed by Analysis of Adversary Campaigns and Intrusion Kill Chains," Bethesda, Md.: Lockheed Martin Corporation, 2011.

"An Information Security Preparedness Tool to Do Adversarial Simulation," webpage, GitHub, last updated June 19, 2018. As of January 8, 2020:
https://github.com/uber-common/metta

InfoSec Institute, "CIA Triad," webpage, February 7, 2018. As of January 4, 2020:
https://resources.infosecinstitute.com/cia-triad/

INSA—*See* Intelligence and National Security Alliance.

Intelligence and National Security Alliance, *A Framework for Cyber Indications and Warning*, Arlington, Va., October 2018.

Intelligence and National Security Alliance, Cyber Intelligence Task Force, *Operational Levels of Cyber Intelligence*, September 2013.

Joint Publication 2-0, *Joint Intelligence*, Washington, D.C.: Joint Chiefs of Staff, October 22, 2013.

Joint Publication 3-0, *Joint Operations*, Washington, D.C.: Joint Chiefs of Staff, Incorporating Change 1, October 22, 2018.

Joint Publication 3-13, *Information Operations*, Washington, D.C.: Joint Chiefs of Staff, Incorporating Change 1, November 20, 2014.

Khalsa, Sundri, "The Intelligence Community Debate over Intuition Versus Structured Technique: Implications for Improving Intelligence Warning," *Journal of Conflict Studies*, Vol. 29, April 1, 2009.

Knake, Robert K., "Building Resilience in the Fifth Domain," blog post, Council on Foreign Relations, July 16, 2019. As of January 4, 2020:
https://www.cfr.org/blog/building-resilience-fifth-domain

Ko, Leekyung, "North Korea as a Geopolitical and Cyber Actor: A Timeline of Events," blog post, New America, June 6, 2018. As of January 4, 2020:
https://www.newamerica.org/cybersecurity-initiative/c2b/c2b-log/
north-korea-geopolitical-cyber-incidents-timeline/

Kong, Ji Young, Jong In Lim, and Kyoung Gon Kim, "The All-Purpose Sword: North Korea's Cyber Operations and Strategies," in Tomáš Minárik, Siim Alatalu, Stefano Biondi, Massimiliano Signoretti, Ihsan Tolga, and Gábor Visky, eds., *11th International Conference on Cyber Conflict: Silent Battle, Proceedings 2019*, Tallinn: NATO Cooperative Cyber Defence Centre of Excellence, 2019.

Lakhno, Valeriy, Svitlana Kazmirchuk, Yulia Kovalenko, Larisa Myrutenko, and Tetyana Okhrimenko, "Design of Adaptive System of Detection of Cyber-Attacks, Based on the Model of Logical Procedures and the Coverage Matrices of Features," *East European Journal of Advanced Technology*, Vol. 3, No. 9, June 2016, pp. 30–38.

Langeland, Krista S., David Manheim, Gary McLeod, and George Nacouzi, *How Civil Institutions Build Resilience: Organizational Practices Derived from Academic Literature and Case Studies*, Santa Monica, Calif.: RAND Corporation, RR-1246-AF, 2016. As of January 4, 2020:
https://www.rand.org/pubs/research_reports/RR1246.html

Lee, Robert M., Michael J. Assante, and Tim Conway, *Analysis of the Cyber Attack on the Ukrainian Power Grid*, Washington, D.C.: SANS Industrial Control Systems, Electricity Information Sharing and Analysis Center, March 18, 2016.

Lilly, Bilyana, Lillian Ablon, Quentin Hodgson, and Adam Moore, "Applying Indications and Warning Frameworks to Cyber Incidents," in Tomáš Minárik, Siim Alatalu, Stefano Biondi, Massimiliano Signoretti, Ihsan Tolga, and Gábor Visky, eds., *11th International Conference on Cyber Conflict: Silent Battle, Proceedings 2019*, Tallinn: NATO Cooperative Cyber Defence Centre of Excellence, 2019.

Lyngaas, Sean, "UN Report Links North Korean Hackers to Theft of $571 Million from Cryptocurrency Exchanges," *CyberScoop*, March 12, 2019. As of January 4, 2020:
https://www.cyberscoop.com/un-report-accuses-north-korean-hackers-stealing-571-million-crytocurrency-exchanges/

Maimon, David, Olga Babko-Malaya, Rebecca Cathey, and Steve Hinton, "Re-Thinking Online Offenders' SKRAM: Individual Traits and Situational Motivations as Additional Risk Factors for Predicting Cyber Attacks," *IEEE 15th International Conference on Dependable, Autonomic and Secure Computing, 15th International Conference on Pervasive Intelligence and Computing, 3rd International Conference on Big Data Intelligence and Computing, and Cyber Science and Technology Congress (DASC/PiCom/DataCom/CyberSciTech)*, Orlando, Fla., November 6–10, 2017.

Maltego, "Maltego Products," webpage, undated. As of September 14, 2020:
https://www.maltego.com/products/?utm_source=paterva.
com&utm_medium=referral&utm_campaign=301

Mathew, Sunu, Daniel Britt, Richard Giomundo, and Shambhu Upadhyaya, "Real-Time Multistage Attack Awareness Through Enhanced Intrusion Alert Clustering," *MILCOM 2005 - 2005 IEEE Military Communications Conference*, Atlantic City, N.J., October 17–20, 2005.

Merriam-Webster, "Organization," webpage, last updated September 9, 2020. As of September 14, 2020:
https://www.merriam-webster.com/dictionary/organization

"MISP (Core Software)—Open Source Threat Intelligence and Sharing Platform (Formerly Known as Malware Information Sharing Platform)," webpage, GitHub, undated. As of January 4, 2020:
https://github.com/MISP/MISP

MITRE | ATT&CK, "Adversary Emulation Plans," webpage, undated a. As of January 4, 2020:
https://attack.mitre.org/resources/adversary-emulation-plans/

MITRE | ATT&CK, "Groups," webpage, undated b. As of January 4, 2020:
https://attack.mitre.org/groups/

MITRE | ATT&CK, "PRE-ATT&CK Introduction," webpage, undated c. As of January 4, 2020:
https://attack.mitre.org/resources/pre-introduction/

MITRE | ATT&CK, "Acquire OSINT Data Sets and Information," webpage, last modified October 17, 2018a. As of September 21, 2020:
https://attack.mitre.org/techniques/T1247/

MITRE | ATT&CK, "Buy Domain Name, Procedure Examples," webpage, last modified October 17, 2018b. As of January 4, 2020:
https://attack.mitre.org/techniques/T1328/

MITRE | ATT&CK, "Establish and Maintain Infrastructure, Techniques," webpage, last modified October 17, 2018c. As of January 4, 2020:
https://attack.mitre.org/tactics/TA0022/

MITRE | ATT&CK, "Test Capabilities, Techniques," webpage, October 17, 2018d. As of January 4, 2020:
https://attack.mitre.org/tactics/TA0025/

MITRE | ATT&CK, "Test Malware to Evade Detection," webpage, last modified October 17, 2018e. As of January 4, 2020:
https://attack.mitre.org/techniques/T1359/

MITRE | ATT&CK, "Scripting," webpage, last modified June 24, 2019a. As of January 9, 2020:
https://attack.mitre.org/techniques/T1064/

MITRE | ATT&CK, "Discovery," webpage, last modified July 19, 2019b. As of January 9, 2020:
https://attack.mitre.org/tactics/TA0007/

MITRE | ATT&CK, "Execution," webpage, last modified July 19, 2019c. As of January 9, 2020:
https://attack.mitre.org/tactics/TA0002/

MITRE | ATT&CK, "PRE-ATT&CK Matrix," webpage, last modified November 4, 2019d. As of January 4, 2020:
https://attack.mitre.org/matrices/pre/

MITRE | ATT&CK, "Boot or Logon Autostart Execution: Registry Run Keys/Startup Folder," webpage, last modified March 25, 2020a. As of January 9, 2020:
https://attack.mitre.org/techniques/T1060/

MITRE | ATT&CK, "System Information Discovery," webpage, last modified March 26, 2020b. As of January 9, 2020:
https://attack.mitre.org/techniques/T1082/

MITRE | ATT&CK, "Web Service," webpage, last modified March 26, 2020c. As of January 10, 2020:
https://attack.mitre.org/techniques/T1102/

MITRE | ATT&CK, "Phishing: Spearphishing Attachment," webpage, last modified March 27, 2020d. As of January 10, 2020:
https://attack.mitre.org/techniques/T1193/

MITRE | ATT&CK, "Hijack Execution Flow: Service Registry Permissions Weakness," webpage, last modified June 20, 2020e. As of January 10, 2020:
https://attack.mitre.org/techniques/T1058/

MITRE | ATT&CK, "Signed Binary Proxy Execution: Mshta," webpage, last modified June 20, 2020f. As of January 9, 2020:
https://attack.mitre.org/techniques/T1170/

MITRE | ATT&CK, "Enterprise Matrix," webpage, last modified July 2, 2020g. As of January 4, 2020:
https://attack.mitre.org/matrices/enterprise/

MITRE | ATT&CK, "Updates," webpage, last modified April 29, 2021. As of April 30, 2021:
https://attack.mitre.org/resources/updates/

MITRE Corporation, "MITRE ATT&CK Navigator," webpage, undated. As of January 4, 2020:
https://mitre-attack.github.io/attack-navigator/

Moteff, John D., *Critical Infrastructures: Background, Policy, and Implementation*, Washington, D.C.: Congressional Research Service, RL30153, June 10, 2015.

Moteff, John, and Paul Parfomak, *Critical Infrastructure and Key Assets: Definition and Identification*, Washington, D.C.: Congressional Research Service, RL32631, October 1, 2004.

Muckin, Michael, and Scott C. Fitch, *A Threat-Driven Approach to Cyber Security: Methodologies, Practices and Tools to Enable a Functionally Integrated Cyber Security Organization*, Bethesda, Md.: Lockheed Martin Corporation, 2019.

Nakashima, Ellen, "Russian Spies Hacked the Olympics and Tried to Make It Look Like North Korea Did It, U.S. Officials Say," *Washington Post*, February 24, 2018.

National Infrastructure Advisory Council, *Critical Infrastructure Resilience: Final Report and Recommendations*, Washington, D.C.: U.S. Department of Homeland Security, September 8, 2009.

National Institute of Standards and Technology, *Security and Privacy Controls for Federal Information Systems and Organizations*, Gaithersburg, Md.: Joint Task Force Transformation Initiative, Special Publication 800-53, Revision 4, April 2013. As of January 4, 2020:
https://nvlpubs.nist.gov/nistpubs/SpecialPublications/NIST.SP.800-53r4.pdf

NIST—*See* National Institute of Standards and Technology.

Obama, Barack H., Executive Order 13636, *Improving Critical Infrastructure Cybersecurity*, Washington, D.C.: White House, February 12, 2013. As of January 4, 2020: https://obamawhitehouse.archives.gov/the-press-office/2013/02/12/executive-order-improving-critical-infrastructure-cybersecurity

Oehmen, Christopher S., Paul J. Bruillard, Brett D. Matzke, Aaron R. Phillips, Keith T. Star, Jeffrey L. Jensen, Doug Nordwall, Seth Thompson, and Elena S. Peterson, "LINEBACKER: LINE-Speed Bio-Inspired Analysis and Characterization for Event Recognition," *IEEE Security and Privacy Workshops (SPW)*, San Jose, Calif., May 22–26, 2016, pp. 88–95.

Okutan, Ahmet, Shanchieh Jay Yang, and Katie McConky, "Predicting Cyber Attacks with Bayesian Networks Using Unconventional Signals," *CISRC '17 Proceedings of the 12th Annual Conference on Cyber and Information Security Research*, No. 13, April 2017.

O'Rourke, T. D., "Critical Infrastructure, Interdependencies, and Resilience," *The Bridge*, Vol. 37, No. 1, Spring 2007, pp. 22–29.

Pace, Chris, ed., *The Threat Intelligence Handbook: A Practical Guide for Security Teams to Unlocking the Power of Intelligence*, Annapolis, Md.: CyberEdge Group, 2018.

Park, Ju-min, and James Pearson, "In North Korea, Hackers Are a Handpicked, Pampered Elite," *Reuters*, December 5, 2014.

Pokorny, Zane, "What Is Threat Intelligence? Definition and Examples," blog post, Recorded Future, April 30, 2019. As of January 4, 2020: https://www.recordedfuture.com/threat-intelligence-definition/

Pope, Simon, and Audun Jøsang, "Analysis of Competing Hypotheses Using Subjective Logic," *10th International Command and Control Research and Technology Symposium: The Future of C2, Decision Making and Cognitive Analysis*, January 2005. As of January 4, 2020: https://apps.dtic.mil/dtic/tr/fulltext/u2/a463907.pdf

"A Powershell Incident Response Framework," webpage, GitHub, last updated on August 4, 2020. As of January 4, 2020: https://github.com/davehull/Kansa

President's Commission on Critical Infrastructure Protection, *Critical Foundations: Protecting America's Critical Infrastructures*, Washington, D.C., October 13, 1997.

"Proactive Defense: Understanding the 4 Main Threat Actor Types," Recorded Future, blog post, August 23, 2016.

Quinn, Christian, "The Emerging Cyberthreat: Cybersecurity for Law Enforcement," *Police Chief*, December 12, 2018. As of January 5, 2020: https://www.policechiefmagazine.org/the-emerging-cyberthreat-cybersecurity/

RiskIQ, "RiskIQ Community Edition," web tool, undated. As of January 4, 2020: https://www.riskiq.com/products/community-edition/

Robb, Drew, "Eight Top Threat Intelligence Platforms," *eSecurity Planet*, July 18, 2017.

Robinson, Michael, Craig Astrich, and Scott Swanson, "Cyber Threat Indications and Warning: Predict, Identify and Counter," *Small Wars Journal*, July 26, 2012.

"Scalable Automated Adversary Emulation Platform," webpage, GitHub, last updated January 8, 2020. As of January 8, 2020: https://github.com/mitre/caldera

Sharma, Anup, Robin Gandhi, Qiuming Zhu, William Mahoney, and William Sousan, "A Social Dimensional Cyber Threat Model with Formal Concept Analysis and Fact-Proposition Inference," University of Nebraska Omaha, Computer Science Faculty Publications, No. 24, December 2013.

Singh, Jai, "The Lockwood Analytical Method for Prediction Within a Probabilistic Framework," *Journal of Strategic Security*, Vol. 6, No. 3, Fall 2013, pp. 83–99.

Slowik, Joe, "Threat Analytics and Activity Groups," blog post, *Dragos*, February 26, 2018. As of June 23, 2020:
https://www.dragos.com/blog/industry-news/threat-analytics-and-activity-groups/

"Small and Highly Portable Detection Tests Based on MITRE's ATT&CK," webpage, GitHub, last updated January 8, 2020. As of January 8, 2020:
https://github.com/redcanaryco/atomic-red-team

Strom, Blake, "ATT&CK 101," blog post, Medium, May 3, 2018. As of January 4, 2020:
https://medium.com/mitre-attack/att-ck-101-17074d3bc62

Taft, J. D., *Electric Grid Resilience and Reliability for Grid Architecture*, Richland, Wash.: Pacific Northwest National Laboratory, U.S. Department of Energy, PNNL-26623, November 2017. As of January 4, 2020:
https://gridarchitecture.pnnl.gov/media/advanced/Electric_Grid_Resilience_and_Reliability.pdf

Taylor, Guy, "Signs Point to Renewed Talks Between U.S., North Korea," *Washington Times*, November 10, 2019. As of January 9, 2020:
https://www.washingtontimes.com/news/2019/nov/10/
signs-point-to-renewed-talks-between-us-north-kore/

"A Toolset to Make a System Look as If It Was the Victim of an APT Attack," webpage, GitHub, last updated June 13, 2018. As of January 8, 2020:
https://github.com/NextronSystems/APTSimulator

Trevors, Matthew, "Mapping Cyber Hygiene to the NIST Cybersecurity Framework," webpage, Software Engineering Institute, Carnegie Mellon University, October 30, 2019. As of June 23, 2020:
https://insights.sei.cmu.edu/insider-threat/2019/10/
mapping-cyber-hygiene-to-the-nist-cybersecurity-framework.html

Tunggal, Abi Tyas, "Cyber Resilience: What It Is and Why You Need It," blog post, UpGuard, updated May 18, 2020. As of January 4, 2020:
https://www.upguard.com/blog/cyber-resilience

Unit 42, "Playbook Viewer," webpage, GitHub, undated. As of January 4, 2020:
https://pan-unit42.github.io/playbook_viewer/

United Nations Security Council, "Note by the President of the Security Council," New York, S/2019/171, March 5, 2019. As of January 4, 2020:
https://undocs.org/S/2019/171

U.S. Code of Federal Regulations, Title 48, Chapter 2, Subchapter H, Part 252, Subpart 252.2, Section 252.204-7012, Safeguarding Covered Defense Information and Cyber Incident Reporting.

U.S. Department of the Army, *Intelligence,* Field Manual No. 2-0, Washington, D.C., May 2004.

U.S. Department of Defense, Defense Science Board, *Task Force Report: Resilient Military Systems and the Advanced Cyber Threat*, Washington, D.C.: Office of the Under Secretary of Defense for Acquisition, Technology and Logistics, January 2013.

U.S. Department of Homeland Security, "Statement by Secretary Jeh Johnson on the Designation of Election Infrastructure as a Critical Infrastructure Subsector," Office of the Press Secretary, January 6, 2017. As of January 4, 2020:
https://www.dhs.gov/news/2017/01/06/statement-secretary-johnson-designation-election-infrastructure-critical

U.S. Department of Homeland Security, Cybersecurity and Infrastructure Security Agency, "Critical Infrastructure Sectors," webpage, last revised March 24, 2020. As of January 4, 2020:
https://www.dhs.gov/cisa/critical-infrastructure-sectors

U.S. Marine Corps, "ASCOPE/PMESII," webpage, undated. As of January 4, 2020:
https://www.trngcmd.marines.mil/Portals/207/Docs/wtbn/MCCMOS/Planning%20Templates%20Oct%202017.pdf?ver=2017-10-19-131249-187

"A Utility to Generate Malicious Network Traffic and Evaluate Controls," webpage, GitHub, last updated September 18, 2019. As of January 8, 2020:
https://github.com/alphasoc/flightsim

Watson, Bruce W., Susan M. Watson, and Gerald W. Hopple, eds., *United States Intelligence: An Encyclopedia*, New York: Garland Publishing, 1990.

Watters, Paul A., Stephen McCombie, Robert Layton, and Josef Pieprzyk, "Characterising and Predicting Cyberattacks Using the Cyber Attacker Model Profile (CAMP)," *Journal of Money Laundering Control*, Vol. 15, No. 4, 2012, pp. 430–441.

"Web App That Provides Basic Navigation and Annotation of ATT&CK Matrices," webpage, GitHub, 2018. As of January 4, 2020:
https://github.com/mitre-attack/attack-navigator

Wilson, James Q., *Bureaucracy*, new edition, New York: Basic Books, 2000.

World Economic Forum, *Partnering for Cyber Resilience: Risk and Responsibility in a Hyperconnected World—Principles and Guidelines*, Geneva, Switzerland, 2012.

"Yara in a Nutshell," webpage, GitHub, undated. As of June 23, 2020:
https://virustotal.github.io/yara/

"A Yara Rule Generator for Finding Related Samples and Hunting," webpage, GitHub, last updated November 6, 20 8. As of January 4, 2020:
https://github.com/AlienVault-OTX/yabin

Young, Joseph, "North Korea Hacked Crypto Exchanges and Ran ICOs to Fund Regime: Report," *CCN*, November 13, 2018.